P9-CKX-650

recent advances in phytochemistry

volume 19

Chemically Mediated Interactions between Plants and Other Organisms

RECENT ADVANCES IN PHYTOCHEMISTRY

Proceedings of the Phytochemical Society of North America
General Editor: Eric E. Conn, *University of California, Davis, California*

Recent Volumes in the Series

Volume 11   **The Structure, Biosynthesis, and Degradation of Wood**
Proceedings of the Sixteenth Annual Meeting of the Phytochemical Society
of North America, Vancouver, Canada, August, 1976

Volume 12   **Biochemistry of Plant Phenolics**
Proceedings of the Joint Symposium of the Phytochemical Society of
Europe and the Phytochemical Society of North America, Ghent, Belgium,
August, 1977

Volume 13   **Topics in the Biochemistry of Natural Products**
Proceedings of the First Joint Meeting of the American Society of
Pharmacognosy and the Phytochemical Society of North America,
Stillwater, Oklahoma, August, 1978

Volume 14   **The Resource Potential in Phytochemistry**
Proceedings of the Nineteenth Annual Meeting of the Phytochemical Society
of North America, Dekalb, Illinois, August, 1979

Volume 15   **The Phytochemistry of Cell Recognition**
and Cell Surface Interactions
Proceedings of the First Joint Meeting of the Phytochemical Society of
North America and the American Society of Plant Physiologists,
Pullman, Washington, August, 1980

Volume 16   **Cellular and Subcellular Localization in Plant Metabolisr**
Proceedings of the Twenty-first Annual Meeting of the Phytochemical Socie
of North America, Ithaca, New York, August, 1981

Volume 17   **Mobilization of Reserves in Germination**
Proceedings of the Twenty-second Annual Meeting of the Phytochemical
Society of North America, Ottawa, Ontario, Canada, August, 1982

Volume 18   **Phytochemical Adaptations to Stress**
Proceedings of the Twenty-third Annual Meeting of the
Phytochemical Society of North America, Tucson, Arizona, July, 1983

Volume 19   **Chemically Mediated Interactions between Plants and**
Other Organisms
Proceedings of the Twenty-fourth Annual Meeting of the Phytochemical Soc.
of North America, Boston, Massachusetts, July, 1984

A Continuation Order Plan is available for this series. A continuation order will bring delivery
each new volume immediately upon publication. Volumes are billed only upon actual shipme
For further information please contact the publisher.

recent advances in phytochemistry

volume 19

Chemically Mediated Interactions between Plants and Other Organisms

Edited by

Gillian A. Cooper-Driver
Tony Swain
Boston University
Boston, Massachusetts

and

Eric E. Conn
University of California, Davis
Davis, California

PLENUM PRESS • NEW YORK AND LONDON

Science
QK
861
R29
V.19

Library of Congress Cataloging in Publication Data

Phytochemical Society of North America. Meeting (24th: 1984: Boston, Mass.)
    Chemically mediated interactions between plants and other organisms.
    (Recent advanced in phytochemistry; v. 19)
    "Proceedings of the Twenty-fourth Annual Symposium of the Phytochemical Society of North America on biochemical interactions of plants with other organisms, held July 9–13, 1984, at Boston University, Boston, Massachusetts"— T.p. verso.
    Bibliography: p.
    Includes index.
    1. Botanical chemistry—Congresses. 2. Plant metabolites—Congresses. 3. Botany—Ecology—Congresses. 4. Allelopathic agents—Congresses. I. Cooper-Driver, Gillian A. II. Swain, T. III. Conn, Eric E. IV. Title.
QK861.R38 vol. 19   581.19′2 s   [581.2′3]   85-6419
ISBN 0-306-42006-6

Proceedings of the Twenty-fourth Annual Symposium of the Phytochemical Society of North America on Biochemical Interactions of Plants with Other Organisms, held July 9–13, 1984, at Boston University, Boston, Massachusetts

© 1985 Plenum Press, New York
A Division of Plenum Publishing Corporation
233 Spring Street, New York, N.Y. 10013

All rights reserved

No part of this book may be reproduced, stored in a retrieval system, or transmitted, in any form or by any means, electronic, mechanical, photocopying, microfilming, recording, or otherwise, without written permission from the Publisher

Printed in the United States of America

9/24/85

PREFACE

Chemical warfare between plants and their herbivores
and pathogens was first brought to our attention by the
publication 25 years ago of the paper by Fraenkel in
Science. There, he pointed out that most plants have
similar nutritional characteristics so that the selection
of plants by insect herbivores must depend on the relative
toxicity of secondary compounds. This led, rather gradually,
to a host of papers on plant-herbivore interactions. More
or less at the same time, insect physiologists and ecologists
were starting to realise the importance of chemical communication systems in determining sexual and other characteristics
of insect behaviour.

Nine years ago the Phytochemical Society of North America
published their Symposium on 'Biochemical Interaction Between
Plants and Insects' in which the plant apparency theory was
expounded by both Paul Feeny and Rex Cates and David Rhoades.
This stated that plants which are apparent usually contain
secondary components which reduce digestibility (tannins and
lignins) while ephemeral plants have more toxic, and perhaps
less costly, compounds such as alkaloids. These papers
stimulated much research on biochemical ecology.

The recognition of the importance of the biochemical
factors in such interactions is not just of scientific
interest. It is vitally important in programs for the
production of new varieties of cultivated plants, especially
in tropical countries where about one-third or more of the
crops are lost to predation or disease.

It seemed timely, therefore, for the Society to organize
a Symposium which examined all aspects of the effects of
plant secondary compounds on a whole variety of ecological
interactions. This volume includes nine chapters based on
papers given at a Symposium on 'Biochemical Interactions of
Plants with Other Organisms' during the 24th meeting of the
PSNA held at Boston University in July 1984.

v

Two of the papers deal with the interaction of plants and pathogenic fungi. In this area, there are two distinct problems. How does a plant 'recognize' a potential pathogen? And having done so, how does it 'respond'? The chemical self-not self recognition factors in nature are, as far as we know, almost always glyco- or lipoproteins. Arthur Ayers and his colleagues, in their chapter, describe the nature of these factors and how one can probe the interactions involved with the help of race-specific monoclonal antibodies. This approach is bound to be of great importance in the breeding or genetic engineering of cultivated plants.

Lee Creasy discusses the biochemical responses of plants to attack and how they produce phytoalexins or high molecular-weight compounds to protect themselves. He reviews other current knowledge of the biochemical responses of plants to disease.

Yigal Elad and Iraj Misaghi, in their presentation, perceptively outline the interactions between plants and microorganisms in the soil. This area of research is exciting (and surprisingly rather neglected) since it is obvious that plants need to ensure an adequate supply of minerals, especially nitrogen, from the soil to survive and here are in direct competition with the soil micro-flora and fauna. The rhizosphere is extremely complex. Plants exude toxins, enzymes, metal chelators and so on which need to act effectively if the plant is to survive.

Of course, plants compete with each other in any given ecosystem. This is usually explained as being due to aggressive root growth, shading and variations in the timing of flowering or fruit set. The importance of allelopathy, the effect of chemicals exuded by one plant species on the germination, seedling growth and development of another, is often ignored. Elroy Rice, who has contributed so much to this field, presents a perceptive overview of the subject. The questions of plant succession, patterning of vegetation, geographical variation and the like is dealt with in a masterly manner.

In plant-insect interactions, attention is now being centered on 'multichemical responses' rather than the effect of a single compound present in a host plant. Thus, it is now thought that the whole mixture of secondary compounds

in plants act not only as deterrants to feeding by insects
and other herbivores but also as signals to indicate the
occurrence of desirable foods or to advertise the presence
of host insects for parasitic organisms. This is exemplified
in the chapter by Pedro Barbosa and James Saunders. They
demonstrate that parasites 'home' in on their hosts by
chemical clues given by the plants on which their victims
feed. Again this has wide ranging implications in elimi-
nating the insects of cultivated plants and forest trees
(where the losses from certain insects are enormous). May
Berenbaum illustrates the idea of a 'response spectrum' to
plant chemicals by insects whereby the animal relies on
the presence of many compounds in the host plant to determine its acceptibility as a food or for ovipositing. Isao
Kubo reinforces the ideas of multichemical effects by his
delineation of the resistance of the olive tree to both
fungal and insect attack. It seems likely that for plants,
what is a good protection against pathogens may have
equally deleterious effects against herbivores! Of course
this is what we might expect if we consider the unity of
biochemical mechanisms..

The last two chapters deal with new ideas and challenge
existing theories. David Rhoades provide evidence for
'pheromonal' communication between plants. He describes
how when one plant is devastated by insect herbivores,
nearby plants of the same species respond by increasing
quantitatively their chemical defences. One might expect
that the herbivores may evolve (or already have) a counter
to this strategy and the results he presents are bound to
stimulate research and thought.

The final chapter, by John Bryant, details ongoing
research on the deterrant effect of secondary compounds on
feeding by the arctic snowshoe hare. In a harsh environment,
where food is obviously scarce, it might be expected that
evolution would have provided the hare with detoxification
mechanisms to deal with most of the chemical defences
present in the plants available. But, as is outlined, this
is not the case. These observations argue against the
simplistic optimal foraging hypothesis since there is a
complete lack of correlation between nutrient availability
and plant selection.

Of course, as in all PSNA meetings, there were a selection of excellent short contributed presentations and posters which are not reproduced here. Nevertheless, all the research presented reinforced the Society's aim in demonstrating the importance of secondary metabolism in ecology and evolution. We cannot stress too strongly the feelings of all who attended this Symposium that a better understanding of the biochemical interactions between organisms is an absolute necessity if we are to improve our agricultural and forest productivity and better understand the living world.

We would like to warmly acknowledge the help of all the participants at this Symposium, the Phytochemical Society of North America and Pergammon Press for financial support, the Committee of the PSNA, and Professor Dieter, Dr. R. Buchsbaum, Ms. M. Blewitt, and Mrs. T. Steinharter, all of Boston University, for help in the organization of the Symposium. We also wish to thank Ms. Billie Gabriel for her skillful preparation of the camera-ready copy.

January, 1985   Gillian Cooper-Driver
                                                 Tony Swain
                                                 Eric E. Conn

CONTENTS

Chapter One

PLANT DETECTION OF PATHOGENS

ARTHUR R. AYERS, JODY J. GOODELL
AND PAUL L. DeANGELIS

Cellular and Developmental Biology Department
Harvard University
16 Divinity Avenue
Cambridge, Massachusetts 02138

INTRODUCTION

Plant defense against microorganisms consists of a
variety of preformed barriers and dynamic responses.
Numerous secondary metabolites are present in sufficient
concentrations to be toxic and useful as chemical barriers.
Structural materials, such as the cuticle and cell walls,
are effective as physical barriers to invasion. Preformed
barriers require no plant response to be effective deterrents
to colonization by microorganisms. Plants can, however,
also respond defensively to the presence of some microorganisms. These active responses encompass a complex of

1

strategies to minimize the impact of the invading microorganism. But, active defenses must be triggered to be
effective.

Defensive responses are predicated on detection of
invading microorganisms. The pathogen must therefore
provide the signals that somehow trigger plant responses.
Several classes of molecules derived from pathogens have
been shown to be effective as signals. Some of these signal
molecules act directly, whereas others, such as some enzymes,
act indirectly by converting plant components into signal
molecules.

Plant responses may be very complex, but the responses
to different pathogens are surprisingly similar. Thus,
defense against the many different microorganisms a plant
might encounter is provided not by individualized defenses
to each organism, but rather by a very flexible detection
system that triggers a defense complex of broad effectiveness.

This review is concerned with the identification of
pathogen signal molecules. Fungal pathogens will be the
primary focus, but many of the phenomena described here
will have general application to other microorganisms, and
to animal pests as well.

PLANT RESPONSES TO PATHOGENS

Responses to pathogens range from rapid to very slow
(Fig. 1). Hypersensitive cell death can occur within
seconds after exposure to pathogens, whereas tissue responses, such as development of abscission zones in the petioles
of infected leaves, may require weeks. These responses and
others (Fig. 2) will each be briefly discussed separately,
but they often occur in an interconnected sequence in the
presence of pathogens. In many cases, such as hypersensitive cell death and phytoalexin accumulation, it is difficult
to determine if the responses are independently triggered,
or sequentially expressed in response to a single signal.
Individual defenses have been emphasized as being responsible
for resistance to particular pathogens. It is more likely,
however, that a plant will only be free of the threat of an
invading pathogen after a series of defenses has been
expressed in a synchronized fashion. Thus, even though a

particular response may be essential to resistance, multiple defenses may be necessary to completely neutralize a pathogen.

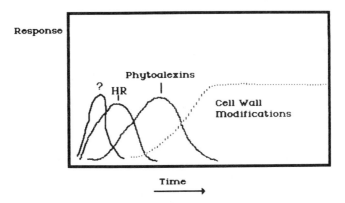

Fig. 1. Timing of plant responses to pathogens.

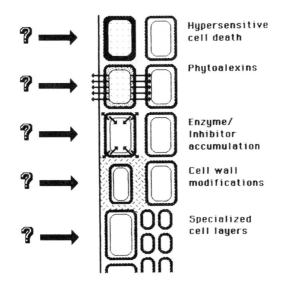

Fig. 2. Responses of plants to pathogens.

Hypersensitive Cell Death

The rapid death of plant cells in the immediate vicinity
of a pathogen has been termed a "hypersensitive" reaction
(HR). The symptoms of HR include granulation of cytoplasm
and loss of organelle integrity. Numerous hypotheses have
been proposed to account for the observed decrease in the
rate of colonization in the vicinity of plant cells responding hypersensitively, but the data are not yet conclusive.[1]
Pathogens requiring living tissue to supply their needs might
be poorly adapted for growth in the environment resulting
from HR. Alternatively, HR may release signals to surrounding cells to stimulate the accumulation of phytoalexins.

Phytoalexin Accumulation

Phytoalexins are toxic secondary products of low molecular weight that accumulate in plant tissues infected by
microorganisms.[2,3] The rate of accumulation of phytoalexins
parallels a reduction in the rate of colonization in some
plant-microorganism interactions.[4,5] As the concentration
of phytoalexins in infected tissue increases to levels that
are inhibitory to the invading microorganism, the invader
ceases to colonize new tissue and is localized.

Inhibition studies provide further support for a central
role of phytoalexins in plant resistance, although they are
complicated by the fact that the enzymes inhibited might also
be important in other facets of the resistance complex.
Inhibition of phytoalexin biosynthesis has been observed to
permit otherwise incompetent organisms to colonize.[6]

Phytoalexin accumulation is controlled at the level of
transcription.[7,8] Messenger RNA's coding for the enzymes
involved in phytoalexin biosynthesis are detectable within
minutes of the addition of an elicitor of phytoalexins.
Within a few hours the enzymes used to synthesize a phytoalexin have been translated and phytoalexins begin to
accumulate. Soybean tissue exposed to an elicitor of
phytoalexins becomes inhibitory to colonization by fungal
pathogens within 6-8 hr after exposure to the elicitor.[9]

The structural genes for some of the enzymes involved
in the synthesis of a parsley phytoalexin have been isolated
and cloned for study.[7]

Enzyme Induction

Plants accumulate a variety of enzymes in response to the presence of microorganisms. Peroxidases and several hydrolytic enzymes are found in increased amounts in infected tissues.[10] These enzymes may attack pathogen structures to impede the invasion or to release signal molecules from the pathogen to improve detection.

Enzyme Inhibitors

Infection or wounding of the lower leaves of a plant, such as tomato, can result in the accumulation of protease inhibitors in the upper leaves.[11] It appears that plants can adapt to predation or infection by interfering with the action of enzymes used to attack their tissues.

Cell Wall Modifications

The primary cell walls of plants are composed of poly-saccharides and proteins that are linked to each other to provide the tensile strength, elasticity and elongation potential characteristic of these structures. Pathogens release enzymes to degrade cell wall components for use of the degradation products as nutrients and to penetrate plant tissues. The cell walls can in turn be modified by the plant to minimize degradability.[12,13] Modifications include lignification,[14] and deposition of hydroxyproline-rich proteins.[15,16]

Modified Cell Layers

Plant tissues can also be modified in response to infection by development of cork layers or through abscission of an infected leaf.

PATHOGEN SIGNAL MOLECULES

The brief description of plant responses presented here and more thoroughly elsewhere in this meeting attests to the versatility of plants. Response is shaped by situation. Thus, a plant must differentiate between types of situations and respond appropriately. Pathogens, bacteria, fungi and viruses signal their presence in different ways. Viruses, for example, do not produce the hydrolytic enzymes used by bacteria and fungi to macerate and digest plant cell walls. On the other hand, viruses present their genome and gene products as potential signals.

Signal molecules can be divided into two groups for convenience, direct and indirect. The direct signals are molecules of the pathogen thought to interact with corresponding plant detection molecules and thereby trigger a response. Indirect signal molecules, on the other hand, are predominantly enzymes thought to generate direct signal molecules by their actions on plant tissue.

Direct signal molecules include common components of fungal cell walls, such as oligosaccharides derived from chitin,[17] chitosan[18] and base-insoluble glucan.[19] Peptides, including the first elicitor of phytoalexins, monilicolin A,[20] and glycoproteins[21,22,23] have also been demonstrated to stimulate plant responses. In each of these cases, the response has been the accumulation of phytoalexins or HR. Phytoalexin accumulation is the best understood of the plant responses to pathogens and may serve as a model for other responses.

Indirect signal molecules are typified by fungal[24] or bacterial[25] enzymes that degrade plant cell wall pectin (Fig. 3). The oligosaccharides resulting from pectin degradation are complex in structure and active as direct elicitors of phytoalexin accumulation. Some of these oligosaccharides are also able to stimulate localized accumulation of hydroxyproline-rich proteins[16] and systemic accumulation of protease inhibitors.[11] The capacity of plants to respond to fragments of plant cell walls provides a new approach to understanding plant phenomena.[26]

One direct pathogen signal molecule, the fungal wall glucan that elicits phytoalexins, will be described in more detail as an example to illustrate some of the approaches now used to study plant detection of pathogens.

GLUCAN ELICITOR OF PHYTOALEXINS

A prominant component of the walls of many genera of fungi is a base-insoluble glucan composed of β-1,3-linked glucosyl residues as linear polymers that are in some cases linked to chains of β-1,6-linked glucosyl residues.[19,27,28]

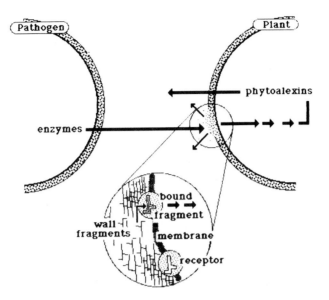

Fig. 3. Plant cell wall fragments released by pathogen
enzymes bind to plant membrane receptor to trigger phytoalexin
accumulation.

Cultures of fungi with cell walls containing this glucan,
such as Phytophthora megasperma f.sp. glycinea, spontaneously
release into the culture medium glucan oligosaccharides that
elicit phytoalexin accumulation in plant tissues.[27] These
signal molecules have been called the glucan "elicitor".

     Glucan elicitor can be obtained directly from culture
filtrates[27] or released by hot water extraction,[28] acid
hydrolysis,[9] or enzymatic degradation of purified fungal
walls. The resulting active oligosaccharides range in size
from > 200 kd to approximately 1200 d. The smallest oligo-
saccharide with elicitor activity (Fig. 4) is a glucan with
five 1,6-linked glucosyl residues and two additional non-
reducing terminal glucosyl residues attached to carbon-3 of
the second and fourth of the five-membered backbone.[19] This
heptamer has also been synthesized and its activity as an
elicitor confirmed.[29] Of the hundreds of possible glucan
heptamers, this is the only isomer with phytoalexin elicitor
activity. Less than 20 ng was the half saturation level for

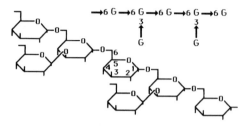

Fig. 4. Structure of heptamer elicitor, the smallest glucan oligomer known to be active in stimulating phytoalexin accumulation.

phytoalexin accumulation in soybean cotyledons.[19] This represents an amazing example of the specificity and sensitivity with which plants can recognize molecules!

The minimal heptamer structure of the glucan elicitor indicates that terminal glucosyl residues with a specific spatial orientation are required for elicitation of phytoalexins.[19] These studies do not, however, give a clear indication of the structures responsible for elicitor activity in the larger size classes. Several alternatives are possible. The larger glucan oligosaccharides may be degraded in plant tissue to yield the active heptamer. Alternatively, the longer chains of linear 3-linked glucosyl residues of the larger oligosaccharides may provide terminal glucosyl residues with more flexible orientations. It is likely that as the side chains become longer, the constraints on the configuration of the backbone to which they are attached will be relaxed. Thus, glucans may be active without having the heptamer structure. In fact, the examination of larger oligosaccharides may reveal many different structures that all satisfy the same stereochemical constraints necessary for activity.

The structures required for elicitor activity in larger glucan oligomers could be examined by extensions of the analytical and synthetic techniques used to study the heptamer. Unfortunately, the complexity of this approach prohibits its general use. Even the heptamer will not be made available to the scientific community (P. Albersheim, personal communication). There are, however, alternative approaches which we have developed in our laboratory.

Linear oligosaccharides of laminarin (β-1,3-linked glucan), can be activated by reductive amination in the presence of cyanoborohydride and an appropriate linker molecule containing a primary or secondary amine.[30] With proper choice of the linker, the oligosaccharides can be linked in dimers or provided with a convenient handle for the incorporation of a variety of labels (radioactive, fluorescent, photoaffinity, etc.).[31] None of the dimers formed to date by linking glucan oligomers (di- and trisaccharides) have been found to have elicitor activity. We have, however, demonstrated that the derivatization process is feasible with natural glucan oligomers and elicitor activity is retained.[31]

We are also pursuing the possibility of producing monoclonal antibodies specific for molecules with elicitor activity. Purified elicitor linked to a protein has been used to demonstrate that antibodies specific for carbohydrates present in the elicitor preparation can be produced as polyclonals in challenged animals. Monoclonals produced from mice challenged with immobilized purified elicitor are now being studied. Elicitor-specific monoclonal antibodies will provide a facile means of purifying molecules with elicitor activity and will provide a very powerful tool in understanding elicitor recognition by plants.

PLANT DETECTION OF SIGNAL MOLECULES -- RECEPTORS

Detection of elicitors is likely to be mediated by binding of the elicitor to receptors present on plant cytoplasmic membranes. Indirect evidence suggests that elicitor receptors exist. Glucan elicitor (or at least preparations rich in glucan elicitor) from Phytophthora infestans is able to agglutinate[32] and kill[32,33] potato protoplasts. Haptens (methyl-glucoside, laminarin and its oligomers, including those found in fungi[33]) of the glucan elicitor produced by P. infestans can effectively block the symptoms of elicitor action on potato tuber tissue[34] and on protoplasts.[32,33] There is also evidence[35] that hyphae of P. infestans bind to the cytoplasmic membranes of infected potato cells. These observations are all consistent with binding of glucan elicitor to cytoplasmic membranes of potato cells. The hapten competition studies indicate that the binding is specific for particular glycosyl residues.

Recently, Yoshikawa, Keen and Wang[36] reported evidence for soybean membrane receptors that bound components of a glucan preparation (mycolaminarin) from a fungal plant pathogen. Mycolaminarin, a storage form of β-1,3-linear glucans with some branches from the 6-position, was radio-labeled by growth on a medium supplemented with [^{14}C] glucose. The specific radioactivity of the mycolaminarin preparation (0.8 mCi/mmole) required very large samples of membranes (30 mg) for binding studies, but the data indicate that membrane receptors are present. They observed binding properties characteristic of a ligand-receptor interaction: pH dependence; rapid saturation kinetics of binding; and temperature dependence. Binding was greatly inhibited both by pretreating the membranes at 60°C for 10 minutes and by addition of proteolytic enzymes, indicating the receptors are heat-labile proteins.

The availability of pure, synthetic elicitors of known structures and with specific activities of radiolabeling appropriate for binding studies will greatly facilitate receptor studies. Pending these results, membrane-bound elicitor receptors must still be considered hypothetical. The distinct possibility still remains that the observed binding represents an interesting artefact, such as the binding of the elicitor as a substrate for hydrolytic enzymes or glucan synthetase. Glucan synthetase, for example, is known to exist on the surface of plasma membrane vesicles[37] and could account for the observed binding. One indication of the potential interaction of elicitor with the glucan synthetase is the recent observation in our laboratory that addition of several glucan oligomers with structures related to the glucan elicitor alters the activity of glucan synthe-tase in purified plasma membrane preparations from soybean roots.

PLANT-PATHOGEN INTERACTIONS

Numerous studies of the genetics of disease resistance in plants have established a pattern in the relationships of hosts and their pathogens. Plants with major genes for resistance exhibit incompatible (pathogen growth restricted) interactions with races of a pathogen with corresponding genes for avirulence (Fig. 5). For example, a variety with resistance genes R_1 and R_2 would actively inhibit the spread of all races with either complementary avirulence genes 1 or

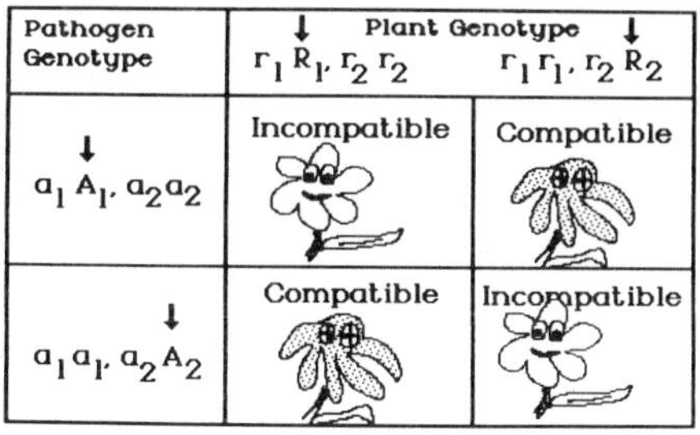

Pathogen Genotype	↓ Plant Genotype ↓ $r_1 R_{1'} r_2 r_2$	$r_1 r_{1'} r_2 R_2$
↓ $a_1 A_{1'} a_2 a_2$	Incompatible	Compatible
↓ $a_1 a_{1'} a_2 A_2$	Compatible	Incompatible

Fig. 5. Interaction of corresponding genes for resistance in plant and avirulence in pathogen. Matching pairs of genes in host and pathogen leads to defensive responses (incompatibility).

2, i.e. race 1, race 2, race 1,2 or race 1,3, but races 3 and 4 would grow and reproduce. If the complementary genes are not expressed in either the host or the pathogen, or both, the pathogen is compatible and grows unimpeded. This same "Gene-for-Gene" correspondence in host and pathogen has been observed in interactions of plants with viruses, bacteria, fungi, nematodes, parasitic higher plants and insects.[38] One obvious molecular interpretation of this pattern is that the plant produces receptors specific for pathogen components. Recognition of pathogen components then triggers stereotyped defense responses that ultimately stop the spread of the pathogen in host tissue.

The glucan elicitor and its putative receptor may be a suitable model for the interactions of plant and pathogen molecules involved in race-specific disease resistance, but these components and the responses must be considered as secondary to the race-specific interactions. The responses are not distinct for each race of pathogen, and neither are the elicitors.[39] The identification of the products of plant resistance genes and pathogen avirulence genes remains a major unanswered problem of biology.

AVIRULENCE GENE PRODUCTS

The genetic characteristics of avirulence and resistance genes have been recently reviewed as a guide to the nature of the products of these genes.[1] Some important

generalizations result from comparisons of host-pathogen
systems. Resistance genes are often found in closely linked
clusters with true allelism, and are inherited as dominant
traits. The corresponding avirulence genes, while dominant,
are generally not clustered and do not exhibit allelism.
Resistance genes may be functional against races of different
pathogens. Temperature sensitive resistance genes have been
identified.

The observed characteristics of genes directly involved
in race-specific disease resistance provide little insight
into the structure of the plant and pathogen components
involved. It would come as a surprise to very few if the
primary functional products of these genes are proteins.
Disease resistance gene products are likely to be receptors
present on plasma membranes of plant cells. The identity of
the pathogen molecules involved in race-specific resistance
is more problematical. Likely candidates are proteins or
carbohydrates. Available evidence provides no means of
discriminating between the alternatives. The existence of
at least one case in which two avirulence genes are paired
with one resistance gene could be reconciled by the avirulence genes coding for two glycosyl transferases used in the
synthesis of a carbohydrate structure recognized by the
resistance gene receptor.[1]

One of the possible models for the interaction of
avirulence and resistance gene products is as follows:

1. Resistance genes code for plasma membrane
   receptors,

2. Avirulence genes code for glycosyl transferases
   used in the synthesis of the glycomoieties of
   extracellular glycoproteins, such as hydrolytic
   enzymes (Fig. 6),

3. The glycosyl residues added by each transferase
   mediate binding to the corresponding receptor,

4. Binding of complementary pathogen components by
   the membrane receptors initiates a series of
   defensive responses leading to resistance.

One of the interesting speculative consequences of this
model is that resistance gene products (receptors) and

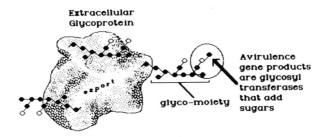

Fig. 6. Avirulence genes of the pathogen are hypothesized
to code for glycosyl transferases that add glycomoieties
to proteins that are subsequently secreted as glycoproteins.

avirulence gene products (glycosyl transferases) have many
common properties. Both are likely to be membrane bound
and have the possibility of being exposed on the plasma
membrane surface. Both bind the glycosyl moieties of glycoproteins. It would be very convenient for a plant to have
a means of sampling the genome of its pathogens and modifying
avirulence genes into corresponding resistance genes for
incorporation into its germ line. Such a system would
provide a direct means by which a plant could obtain new
resistance genes without modifying other genes in its
genome.

IMMUNOCHEMICAL APPROACHES TO AVIRULENCE GENE FUNCTION

    None of the pathogen components related to avirulence
gene products and involved directly in pathogen recognition
has yet been identified. A comprehensive characterization
of pathogen components has been initiated in our laboratory
in an effort to identify these critical race-specific signal
molecules. A major project in this study is the analysis
of the antigens produced by different races of the fungal
pathogen P. megasperma f.sp. glycinea.

    Monoclonal antibodies produced by hybridomas derived
from spleen cells of mice challenged with pathogen extracellular glycoproteins have been screened for their ability
to bind to glycoproteins from different races (Fig. 7).

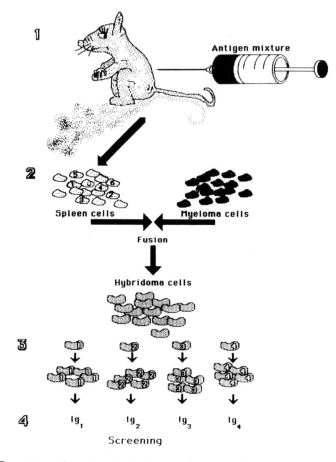

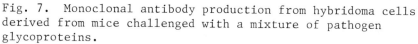

Fig. 7. Monoclonal antibody production from hybridoma cells derived from mice challenged with a mixture of pathogen glycoproteins.

Antibodies from more than one hundred hybridoma colonies have been identified as specific for extracellular glyco-proteins.[31] Four classes of antibodies have been observed (Fig. 8). Class I antibodies bind to the same glycoprotein found in different races and are thought to be specific for the protein portion of the antigen. Class II antibodies

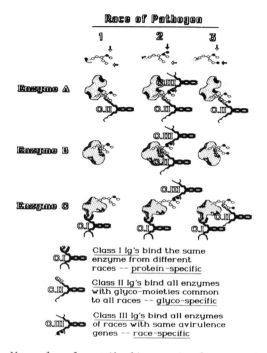

Fig. 8. Monoclonal antibodies raised to a mixture of
glycoproteins can be classified by the pattern of their
interactions with individual glycoproteins.

bind universally to many glycoproteins from each race and
are expected to be specific for the common, backbone portion
of the glycomoieties. Race-specific antibodies of Class III
are recognized by the ability to bind to several different
glycoproteins (in practice extracellular enzymes are used)
of the same race, but not to those of other races. Class
III antibodies are thought to be specific for the race-
specific portion of the glycomoieties. Another class (IV)
of antibodies consists of those that bind to single antigens
present in only one race. Class IV antibodies bind to
antigens that are race-specific or more appropriately,
race-associated, and are likely to be proteins. Comparison
of different isolates of the same race, but from different
geographic locations is likely to show that these antigens
are only coincidentally associated with a particular avirulence gene.

Monoclonal antibodies are expected to permit the identification of race-specific antigens and to facilitate the subsequent identification of resistance gene products as their receptors.

CONCLUSION

The chain of events from the juxtaposition of pathogen spore and plant tissue to the establishment of permanent barriers that prohibit pathogen growth and development remain a mystery at the molecular level. Essential plant and pathogen molecules involved in the early recognition events have not yet been identified. Clearly, plant disease resistance is still a formidable challenge to our attempts to understand plant function.

On the practical side, disease resistance is a central element of crop protection. It seems obvious that more comprehensive and deliberate use of disease resistance would help substantially in reducing dependence on potentially harmful pesticides. With the powerful techniques currently available we expect to be able to identify the products of avirulence genes within a few years. Identification of avirulence and resistance genes will follow shortly thereafter. The availability of cloned resistance genes of known function will provide a powerful new application of genetic engineering of plants to create elite crop varieties with more effective and stable disease resistance.

ACKNOWLEDGMENTS

This work was supported by grants from The National Science (PCM 83-02789) and Maria Moors Cabot Foundations.

REFERENCES

1. KEEN, N.T. 1982. Specific recognition in gene-for-gene host-parasite systems. Adv. Plant Pathol. 1: 35-82.
2. PAXTON, J.D. 1981. Phytoalexins--a working definition. Phytopathol. Z. 101: 106-109.
3. BAILEY, J.A., J.W. MANSFIELD, eds. 1982. Phytoalexins. Blackie, Glasgow/London.

4. BAILEY, J.A. 1974. The relationship between symptom
    expression and phytoalexin concentration in
    hypocotyls of Phaseolus vulgaris infected with
    Colletotrichum lindemuthianum. Physiol. Plant
    Pathol. 4: 477-488.
5. BAILEY, J.A., P.M. ROWELL, G.M. ARNOLD. 1980. The
    temporal relationship between host cell death,
    phytoalexin accumulation and fungal inhibition
    during hypersensitive reactions of Phaseolus
    vulgaris to Colletotrichum lindemuthianum. Physiol.
    Plant Pathol. 17: 329-339.
6. MOESTA, P., H. GRISEBACH. 1982. L-2-Aminoxy-3-
    phenylpropionic acid inhibits phytoalexin accumu-
    lation in soybean with concomitant loss of
    resistance against Phytophthora megasperma f.sp.
    glycinea. Physiol. Plant Pathol. 21: 65-70.
7. HAHLBROCK, K., J. CHAPPELL, D.N. KUHN, M. WALTER,
    E. SCHMELZER. 1984. Induction of resistance-
    related mRNAs by UV light or fungal elicitor in
    cultured plant cells. In Cellular and Molecular
    Biology of Plant Stress, UCLA Symposia on Molecular
    and Cellular Biology, New Series. (J.L. Key, T.
    Kosuge, eds.), Vol. 22, Alan R. Liss, Inc., New
    York, New York (in press).
8. EBEL, J., W.E. SCHMIDT, R. LOYAL. 1984. Phytoalexin
    synthesis in soybean cells: elicitor induction of
    phenylalanine ammonia-lyase and chalcone synthase
    mRNAs and correlation with phytoalexin accumula-
    tion. Arch. Biochem. Biophys. 232: 240-248.
9. ALBERSHEIM, P., B.S. VALENT. 1978. Host-pathogen
    interactions in plants. Plants when exposed to
    oligosaccharides of fungal origin, defend them-
    selves by accumulating antibiotics. J. Cell Biol.
    78: 627-643.
10. BOLLER, T. 1984. Induction of hydrolases as a defense
    reaction against pathogens. In J.L. Key, T. Kosuge,
    eds., op. cit. Reference 7 (in press).
11. RYAN, C.A. 1984. Pectic fragments regulate the
    expression of proteinase inhibitor genes in tomato
    plants. In J.L. Key, T. Kosuge, eds., op. cit.
    Reference 7 (in press).
12. GEBALLE, G.T., A.W. GALSTON. 1982. Wound-induced
    resistance to cellulase in oat leaves. Plant
    Physiol. 70: 781-787.
13. LOEBENSTEIN, G., S. SPIEGEL, A. GERA. 1982. Localized
    resistance and barrier substances. In Active

Defense Mechanisms in Plants. (R.K.S. Wood, ed.),
Plenum Press, New York. pp. 211-230.

14. VANCE, C.P., T.K. KIRK, R.T. SHERWOOD. 1980. Lignification as a mechanism of disease resistance.
Annu. Rev. Phytopathol. 18: 259-288.

15. ESQUERRE-TUGAYE, M.-T., C. LaFITTE, D. MAZAU, A.
TOPPAU, A. TOUZE. 1979. Cell surfaces in plant-
microorganism interactions II, Evidence for the
accumulation of hydroxyproline-rich glycoproteins
in the cell wall of diseased plants as a defense
mechanism. Plant Physiol. 64: 320-326.

16. ESQUERRE-TUGAYE, M.-T., D. MAZAU, B. PELISSIER, D.
ROBY, D. RUMEAU, A. TOPPAN. 1984. Induction by
elicitors and ethylene of proteins associated with
the defense of plants. In J.L. Key, T. Kosuge, eds.,
op. cit. Reference 7 (in press).

17. PEARCE, R.B., J.P. RIDE. 1982. Chitin and related
compounds as elicitors of the lignification response
in wounded wheat leaves. Physiol. Plant Pathol.
20: 199-123.

18. WALTER-SIMMONS, M., L. HADWIGER, C.A. RYAN. 1983.
Chitosans and pectic polysaccharides both induce
the accumulation of the antifungal phytoalexin
pisatin in pea pods and antinutrient proteinase
inhibitors in tomato leaves. Biochem. Biophys.
Res. Commun. 110: 194-199.

19. DARVILL, A.G., P. ALBERSHEIM. 1984. Phytoalexins
and their elicitors -- A defense against microbial
infection in plants. Annu. Rev. Plant Physiol.
35: 243-275.

20. CRUICKSHANK, I.A.M., D.R. PERRIN. 1968. The isolation and partial characterization of monilicolin A,
a polypeptide with phaseollin-inducing activity
from Monilinia fructicola. Life Sci. 7: 449-458.

21. De WIT, P.J.G., E. KODDE. 1981. Further characterization and cultivar-specificity of glycoprotein
elicitors from culture filtrates and cell walls of
Cladosporium fulvum (syn. Fulvia fulva). Physiol.
Plant Pathol. 18: 297-314.

22. De WIT, P.J.G., G. SPIKMAN. 1982. Evidence for the
occurrence of race and cultivar-specific elicitors
of necrosis in intercellular fluids of compatible
interactions of Cladosporium fulvum and tomato.
Physiol. Plant Pathol. 21: 1-11.

23. KEEN, N.T., M. LEGRAND. 1980. Surface glycoproteins. Evidence that they may function as the race-specific phytoalexin elicitors of Phytophthora megasperma f.sp. glycinea. Physiol. Plant Pathol. 17: 175-192.

24. LEE, S.-C., C.A. WEST. 1981. Polygalacturonase from Rhizopus stolonifer, and elicitor of casbene synthetase activity of castor bean (Ricinus communis L.) seedlings. Plant Physiol. 67: 633-639.

25. DAVIS, K.R., G.D. LYON, A. DARVILL, P. ALBERSHEIM. 1982. A polygalacturonic acid lyase isolated from Erwinia carotovora is an elicitor of phytoalexins in soybean. Plant Physiol. 69: 142 (supp.).

26. ALBERSHEIM, P. 1984. Complex carbohydrates regulate stress physiology as well as growth and development. In J.L. Key, T. Kosuge, eds., op. cit. Reference 7 (in press).

27. AYERS, A.R., J. EBEL, F. FINELLI, N. BERGER, P. ALBERSHEIM. 1976. Host-pathogen interactions IX. Quantitative assays of elicitor activity and characterization of the elicitor present in the extracellular medium of cultures of Phytophthora megasperma var. sojae. Plant Physiol. 57: 751-759.

28. AYERS, A.R., J. EBEL, B. VALENT, P. ALBERSHEIM. 1976. Host-pathogen interactions X. Fractionation and biological activity of an elicitor isolated from the mycelial walls of Phytophthora megasperma var. sojae. Plant Physiol. 57: 760-765.

29. OSSOWSKI, P., A. PILOTTI, P. GAREGG, P. LINDBERG. 1983. Synthesis of a branched hepta- and octasaccharide with phytoalexin-elicitor activity. Angew. Chem., Int. Ed. Engl. 22: 793-795.

30. BORCH R.F., M.D. BERNSTEIN, H.D. DURST. 1971. The cyanohydriborate anion as a selective reducing agent. J. Amer. Chem. Soc. 93: 2897-2904.

31. GOODELL, J.J., P. DeANGELIS, A.R. AYERS. 1984. Immunochemical identification of antigens involved in plant/pathogen interactions. In J.L. Key, T. Kosuge, eds., op. cit. Reference 7 (in press).

32. PETERS, B.M., D.H. CRIBBS, D.A. STELZIG. 1978. Agglutination of plant protoplasts by fungal cell wall glucans. Science 201: 364-365.

33. DOKE, N., K. TOMIYAMA. 1980. Effect of hyphal wall components of Phytophthora infestans on protoplasts of potato tuber tissues. Physiol. Plant Pathol. 16: 169-176.

34. MARCAN, H., M.C. JARVIS, J. FRIEND. 1979. Effect of
    methyl glycosides and oligosaccharides on cell
    death and browning of potato tuber discs induced
    by mycelial components of Phytophthora infestans.
    Physiol. Plant Pathol. 14: 1-9.
35. NOZUE, M., K. TOMIYAMA, N. DOKE. 1979. Evidence for
    adherence of host plasmalemma to infecting hyphae
    of both compatible and incompatible races of
    Phytophtora infestans. Physiol. Plant Pathol. 15:
    111-115.
36. YOSHIKAWA, M., N.T. KEEN, M.-C. WANG. 1983. A
    receptor on soybean membranes for a fungal elicitor
    of phytoalexin accumulation. Plant Physiol. 73:
    497-506.
37. HALL, J.L. 1983. Plasma membranes. In Isolation of
    Membranes and Organelles from Plant Cells. (J.L.
    Hall, A.L. Moore, eds.), Academic Press, London,
    New York, pp. 55-81.
38. DAY P.R. 1974. Genetics of Host-Parasite Interac-
    tions, W.H. Freeman, San Francisco, California, 238
    pp.
39. AYERS, A.R., B. VALENT, J. EBEL, P. ALBERSHEIM. 1976.
    Host-pathogen interactions XI. Composition and
    structure of wall released elicitor factions.
    Plant Physiol. 57: 766-774.

Chapter Two

BIOCHEMICAL ASPECTS OF PLANT-MICROBE AND MICROBE-MICROBE
INTERACTIONS IN SOIL

YIGAL ELAD

Department of Plant Pathology and Microbiology
Faculty of Agriculture
The Hebrew University of Jerusalem
Rehovot 76100, Israel

IRAJ J. MISAGHI

Department of Plant Pathology
University of Arizona
Tucson, Arizona 85721

INTRODUCTION

    Beneficial and deleterious microorganisms constantly
interact with each other as well as with plant roots in the
rhizosphere. It is important to elucidate the nature of
microbe-microbe and plant-microbe interactions because plant
health often depends on the outcome of such interactions.

Plants suffer when their roots are attacked by disease-
inducing microbes; they thrive when certain growth promoting
microorganisms manage to colonize their roots. Moreover, it
is becoming evident that the soil and the rhizosphere flora
can be manipulated in ways that favor certain microorganisms
which are capable of reducing the activity of potential plant
pathogens. This approach to disease control, which is known
as biological control, is being actively investigated in
many laboratories.

The outcome of microbe-microbe and plant-microbe
interactions is influenced by many physical, chemical, and
environmental factors. Although parameters that influence
microbe-microbe and plant-microbe interactions can be
defined, their relative importance cannot be easily assessed
in natural ecosystems.

General aspects of biological control of plant pathogens
and biological plant growth promotion have been reviewed.[1-3]
However, little is known about the mechanisms of these two
phenomena. Therefore, this review deals primarily with the
mechanisms of biological control and of biological plant
growth promotion.

MECHANISMS OF BIOLOGICAL CONTROL

Antagonism which plays an important role in biological
control can be due to predation, parasitism, antibiosis,
and competition. An antagonist may reduce the activity,
efficiency, and/or inoculum density of a plant pathogen
and thereby reduce its potential for causing disease.[4,5]
Predation, as observed with mites and vampire amoebae,
currently has not been exploited in biological control[6]
and, therefore, will not be considered here.

Mycoparasitism

Successful biological control of certain soilborne
fungal plant pathogens in the field has been achieved by
infesting soils with certain mycoparasites capable of
parasitizing the pathogens.[7,8] Among mycoparasites,
Trichoderma harzianum and T. hamatum have been studied
extensively. Mycoparasitism begins after mycoparasitic
hyphae make physical contact with the host hyphae. The
contact in some cases may involve recognition. Shortly

after hyphal contact, the mycoparasite may either coil
around the host hyphae or form appressorium-like bodies.[9-11]
Appressoria are sack-shaped structures which are formed by
some fungi above the points of penetration into plant or
fungal cells. Parasitism eventually culminates in the death
of the host cells shortly after they are penetrated by the
mycoparasite. In the following pages the nature of the
events occurring during mycoparasitism will be discussed.

Host-mycoparasite recognition. Recognition has been
shown to be a feature of interactions involving pistil-
pollen, plant-fungi, plant-rhizobia, plant-bacteria,
algae-lichens, and cells of slime molds.[12] Recognition
between fungal host and mycoparasites has only recently been
studied. Elad et al.[13] isolated a hemagglutinin from the
hyphae of Rhizoctonia solani and from the culture filtrate
of the fungus which was capable of agglutinating blood
type O erythrocytes with a high degree of specificity.
The agglutination was dependent on calcium and magnesium
ions and was counteracted by galactose which is present
on the cell wall of the mycoparasite, Trichoderma sp. Elad
and coworkers (unpublished results) later showed that
Sclerotium rolfsii agglutinin agglutinated washed conidia
of Trichoderma sp. They also found that the hyphae of two
host fungi, S. rolfsii and R. solani possess lectins which
bind to carbohydrates on Trichoderma cell walls. The ability
of different isolates of Trichoderma sp. to attack S.
rolfsii was correlated with the ability of S. rolfsii lectin
to agglutinate Trichoderma conidia. D-glucose, D-mannose
and a number of their derivatives, as well as trypsin or
Na_2EDTA, inhibited agglutination, but this inhibition was
reversed by addition of Mn^{2+} and Ca^{2+}. These lectins may
play a major role in recognition.

Induction of multihyphal strand production in the plant
pathogenic fungus, S. rolfsii, by Trichoderma harzianum
seems to involve recognition. Multihyphal strands were
induced following the contact of S. rolfsii with mycelia
of non-parasitizing isolates of T. harzianum, but not with
those of parasitizing isolates.[14] The structure of S.
rolfsii strands was similar to those described for
Phaeococcus exophialae[15] and for Serpula lacrimans.[16]

The development of aggregated structures such as
sclerotia, rhizomorphs, and mycelial strands in certain
fungi also can be induced by interacting microorganisms.

Henis and Inbar[17] reported the induction of microsclerotium
production in R. solani by Bacillus subtilis. The differen-
tiation of Armillaria mellea rhizomorphs also was stimulated
by Aureobasidium pullulans[18] and those of Sphaerostilbe
repens by Penicillium and Aspergillus species.[19] In all
of these cases the development of aggregated structures was
mediated by a diffusable substance. On the other hand, the
stimulation of multihyphal strands in S. rolfsii by T.
harzianum depends on a direct contact between the two fungi.
No substances capable of inducing strands were detected in
culture filtrates of T. harzianum. Moreover, no strands
were formed when colonies of T. harzianum and S. rolfsii
were separated by a membrane. In addition, dead colonies
of T. harzianum did not induce strand formation. The carbon
source on which T. harzianum was grown affected strand
production. Addition of glucose or laminarin to the
cultures prevented the development of strands, whereas
different disaccharides and polysaccharides stimulated
strand formation.[14]

Morphological changes occurring in the fungal host
prior to penetration by mycoparasites. Certain morpho-
gical changes occur in the host hyphae at or near the
host-parasite interface shortly after the contact between
the host and the parasite. The interacting sites of dual
cultures of either R. solani or S. rolfsii and either T.
hamatum or T. harzianum were observed by scanning and
transmission electron microscopes.[9,20] A significant
deposition of intercellular fibrillar materials was observed
outside the interacting cells shortly after a Trichoderma
hypha made contact with the host hyphae. Observations made
in Israel support the idea that there is an increase of a
mucilaginous substance, apparently polysaccharide, originating from either one of the interacting fungi. This was
also found with Stephanoma phaeospora which parasitizes
several species of Fusarium.[21] The significance of these
morphological changes at the interacting sites is not known.
These structures may play a role in recognition or may
constitute a type of defense response against invasion by
mycoparasites. Cell wall modifications which occur in
plants in response to invasion by plant pathogens also have
been suggested as having a role in plant resistance.[22]

Penetration of the host by mycoparasites - the role of
cell-wall degrading enzymes. The composition of cell walls
varies in different classes of fungi. The cell walls of

oomycetes, e.g., Pythium spp. and Phytophthora spp., are
composed of β-glucans, cellulose, and of less than 1.5%
chitin. In contrast, cell walls of the Basidiomycetes and
Ascomycetes are composed of a relatively higher concentration
of chitin as well as β-glucans, and those of the Mucorales
contain chitin and chitosan. Lipids and proteins also are
present in fungal cell walls.[23]

     To parasitize their fungal hosts, mycoparasites must
penetrate cell walls. Although penetration may be aided by
physical means,[9] in most cases cell walls are breached
enzymatically. The various lines of evidence for the enzymatic breakdown of cell walls of host fungi are discussed
below.

     Results of light and electron microscopic studies have
provided visual evidence for the enzymatic degradation of
cell walls of host fungi by a number of mycoparasites,
including Piptocephalis virginiana,[24] Gliocladium virens
which attack Sclerotinia sclerotiorum[25] and Pythium
acanthicum[26] and Verticillium lecanii that parasitize
urediniospores of Puccinia graminis f. sp. tritici.[27]

     Several mycoparasites studied thus far are capable of
producing hydrolytic enzymes such as β-1,3-glucanase,
cellulase, and chitinase which can hydrolyze glucans, cellulose, and chitins, respectively, in the cell wall of fungi.

     The newly discovered[28,29] mycoparasite, Pythium nunn
produces extracellular β-1,3-glucanase, cellulase, and
chitinase when grown on laminarin, cellulose, and chitin,
respectively, as sole carbon sources. The first two enzymes
also were produced when the fungus was grown on the cell
walls of Pythium sp. (N1) and P. aphanidermatum. Large
amounts of β-1,3-glucanase and chitinase were produced by
P. nunn in liquid cultures containing cell walls of R.
solani and S. rolfsii, but low levels were produced in
cultures containing cell walls of Fusarium oxysporum f. sp.
cucumerinum.[30] Moreover, the growth of P. nunn on the cell
wall components of the host fungi was significant when compared to the plant pathogen Pythium sp. (Table 1). The
ability of P. nunn to utilize enzymatic breakdown products
of the cell wall of its hosts also was demonstrated by the
release of $^{14}CO_2$ from labeled cell walls of Pythium sp. (N1)
and R. solani when used as sole food sources[30] (Fig. 1) and
with Trichoderma.[31]

Table 1. The growth and enzymatic activity of the plant pathogen, Pythium sp. (N1) and the mycoparasite, P. nunn in the presence of cell walls of plant pathogens, selected carbohydrates or chitin.

Carbon Source[a]	Growth[b]		Enzymatic Activity of P. nunn[c]		
	Pythium sp. (N1)	P. nunn	β-1,3-Glucanase	Cellulase	Chitinase
Glucose	267	214	0.14	0.23	0.01
Laminarin	310	514	4.74	0.37	0.03
Cellulose	422	319	0.96	3.24	0.25
Chitin	24	116	0.06	0.83	2.17
Rhizoctonia solani cw	134	867	5.09	0.13	1.85
Sclerotium rolfsii cw	67	665	4.16	0.09	1.97
Pythium aphanidermatum cw	200	658	4.54	2.81	0.09
Pythium sp. cw	533	265	4.17	2.72	0.13

[a]Cell wall powder of the host fungi, selected carbohydrates and chitin were furnished at a concentration of 1 mg/ml as the sole carbon source in a synthetic growth medium.

[b]Micrograms of oven dried mycelium/100 ml of synthetic growth medium, after 70 h of incubation at 27°C.

[c]Specific activity was expressed as micromoles of monomer (glucose or N-acetylglucosamine) respectively released in the reaction mixture containing 2 ml of cell free culture, 0.1 M phosphate buffer and 1 mg of laminarin, cellulose, or chitin.

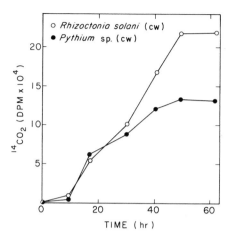

Fig. 1. Release of $^{14}CO_2$ from labeled cell walls of
Rhizoctonia solani and Pythium sp. by P. nunn during 62 h
of incubation. $^{14}CO_2$ was captured on paper discs containing
hyamine (methylbezeldonium) hydroxide and counted in a liquid
scintillation counter. (Elad, Lifshitz and Baker, source).

Although the chitin content of the cell walls of P.
nunn was found to be very low (0.68%), the fungus produced
chitinase when grown on chitin or cell walls containing
chitin. Although Bumbieris[32] reported chitin degradation
by P. ultimum, only P. nunn of all the fungi in the class
Oomycetes tested in our study produced chitinase. Thus,
chitinase production may be required for mycoparasitism by
P. nunn. The pattern of enzyme production by P. nunn in
the presence of fungal cell wall components is similar to
that by Trichoderma sp.[31]

The effects of substrates on enzyme production by
T. harzianum have been studied.[31] Mixtures of laminarin
and glucose in the growth medium varying from 100% laminarin
to 100% glucose and a total carbohydrate concentration of
1.0 mg/ml were tested for their effects on β-1,3-glucanase
activity in vitro. At a carbohydrate concentration of 1.0
mg/ml, the highest enzyme activity (14 glucose units, G.U.
1 unit = 1 micromoles glucose/hr^{-1}mg $^{-1}$ protein) was
obtained with a laminarin to glucose ratio of 3:1 (w/w);
in contrast only 9 and 1.2 G.U. were detected when T.
harzianum was grown on either laminarin alone or glucose

alone, respectively. To determine if glucose serves as a
repressor of β-1,3-glucanase, T. harzianum was grown on
glucose (1.5 mg/ml) for 24 hr before laminarin (0.4 mg/ml)
was added to the medium. The glucose lasted for 72 hr
whereas laminarin was totally utilized within 24 hr after
its addition to the medium. Results show that glucose did
not repress the activity of β-1,3-glucanase at the levels
used in the test.

Chitinase excretion into the growth medium was enhanced
by the presence of chitin at concentrations of 0.5-1.0
mg/ml in the medium. Only slight enzyme activity was
detected when N-acetylglucosamine or glucose were substituted as carbon sources in the medium.

The optimal temperature for excretion of β-1,3-
glucanase by T. harzianum was 30°C and the optimal pH was
5.0. Maximal stability of the enzyme occurred in the pH
range of 7-9. Partial loss of enzyme activity occurred at
35°C and it was completely inactivated at 90°C. β-1,3-
Glucanase exhibited its optimal activity at pH 3-5 and at
40°C.

The optimal temperature and pH for chitinase production
by T. harzianum in the presence of chitin was 28°C at pH 5.
Forty-five and 64% of chitinase activity was lost after
incubation for 1 hr at 40°C and 60°C, respectively, and the
activity was totally lost after 1 hr at 70°C or higher. The
enzyme activity was minimally affected at pH 9. Optimal
activity of chitinase in cell-free extracts was obtained at
35°C and pH 5-6.

These hydrolytic enzymes were detected in dual cultures
of P. nunn and either Phytophthora spp., Mucor sp., Rhizopus
sp., R. solani, S. rolfsii, or one of the several Pythium
spp. but not with ten species of fungi in eight genera in
the class Deuteromycetes or with F. oxysporum f. sp.
cucumerinum.[30]

The mucilaginous layer which coats the cell walls and
macroconidia of F. oxysporum f. sp. cucumerinum[33] may interfere with the induction of cell wall degrading enzymes by
the mycoparasite, P. nunn, by preventing it from making
direct contact with the host cell walls. Cell wall degrading
enzymes of P. nunn released only trace amounts, if any, of
glucose and N-acetyl-D-glucosamine from intact cell walls

of F. oxysporum f. sp. cucumerinum, but substantial amounts
were released when the mucilaginous coating was removed by
treatment with KOH or trypsin. This cell wall component,
which is characteristic of members of Sphaeropsidales and
Tuberculariaceae, may be a factor determining whether cell
wall degrading enzymes are induced or may be involved in
recognition.

Three isolates of T. harzianum with differential
ability to attack S. rolfsii, R. solani, and P. aphanidermatum produced different levels of hydrolytic enzymes. The
levels of enzyme produced in soil were directly correlated
with the ability of these isolates to control soilborne
diseases under greenhouse conditions.[31] However, weak
mycoparasitic isolates of Trichoderma produced a small
amount of lytic enzymes. When the antibiotic actidione
(cycloheximide) was added to a dual culture of S. rolfsii
and T. harzianum, both fungi grew towards each other
normally, but at the meeting zone no parasitism was
observed.[9] Indeed, actidione reduced β-1,3-glucanase and
chitinase activities by 82-83%, respectively. The influence of this antibiotic on other fungal activities which
may be involved in parasitism is not known.

To elucidate the nature and the sequence of events
leading to parasitism, the interactions of P. nunn with a
number of host fungi were studied by light and electron
microscopes. Light microscopy revealed that P. nunn
aggressively parasitizes the following fungi: P. vexans,
P. ultimum, P. aphanidermatum, young hyphae of S. rolfsii
and R. solani, Phytophthora parasitica, P. cinnamomi, P.
citricola, Mucor sp., and Rhizopus sp.[34] P. nunn produced
thin and slender side branches which wrapped around the
surface of host hyphae during the initial stages of parasitism. Late stages of parasitism were of two types termed
"quick reaction" and "slow reaction". The "quick reaction",
noted with the hosts, P. ultimum and P. vexans, was
characterized by massive coiling of the mycoparasite around
the host hyphae. Light microscopy revealed that host
cytoplasm disappeared within a few hours and hyphae frequently burst. The "slow reaction" was noted with hosts
P. aphanidermatum, P. parasitica, P. cinnamomi, P. citricola,
R. solani, S. rolfsii, Mucor sp. and Rhizopus sp. In the
"slow reaction", the mycoparasite wrapped itself only
partially around host hyphae. Globular appressoria ('contact
cells') developed and infection pegs were sometimes formed.

While host cytoplasm disappeared, hyphal bursting and
massive lysis were not observed. Penetration of the host in
the "quick reaction" type of interaction seems to be accomplished more readily than in the "slow reaction". Similarly,
Deacon[35] concluded that coiling of P. oligandrum around the
host fungi may be due to a failure of parasitism and a
reflection of a temporary resistance of the host. Formation
of appressoria was not observed with the mycoparasites P.
acanthicum[26] or P. oligandrum,[36] and according to Barnett
and Binder[37] is not typically found in necrotrophic mycoparasitism. Our observations with P. nunn were consistent
with those of Deacon[35] who demonstrated that young cells
and hyphae of various fungi were generally more prone to be
parasitized by the mycoparasite, P. oligandrum. Similarly,
differences in susceptibility among various oomycete species
to parasitism by P. nunn may be due to variations in properties of the cell wall.

Fluorescent brighteners and fluorescent lectins were
used to study the physical and chemical properties of
surfaces at the sites of interaction between hyphae of the
host and of the mycoparasite. Interacting hyphae of Trichoderma spp. and R. solani or S. rolfsii, and of P. nunn and R.
solani or a pathogenic species of Pythium (N1) were stained
by Calcofluor White M2R New. Although Calcofluor is not a
specific marker, it selectively binds to the edges of polysaccharide oligomers.[38] The appearance of fluorescence
indicated the presence of localized lysis of host cell wall
at points of interaction. A similar phenomenon was observed
with Trichoderma which parasitizes R. solani and S. rolfsii
when walls were stained with a fluorescent lectin.[9] The
specific binding of the fluorescent isothiocyanate (FITC)-
conjugated wheat germ agglutinin to the coiling zones
indicates the presence of N-acetyl-D-glucosamine oligomers
at these sites. Chitin fibrils in the cell walls of S.
rolfsii and R. solani are probably exposed as a result of
the activity of extracellular β-1,3-glucanase excreted by
Trichoderma at contact sites. The presence of D-glucose and
D-mannose in the lysed sites was revealed by concanavalin.
The cell walls of the plant pathogenic fungi which were
tested contained only low concentrations of galactose.
Indeed, no binding of peanut agglutinin was found after the
attack by Trichoderma. Binding of FITC-lectins to the walls
was shown to be sugar specific; no fluorescence was observed
in the presence of both the lectins and their hapten sugars.
These results are in agreement with those of Mirelman et al.[3]

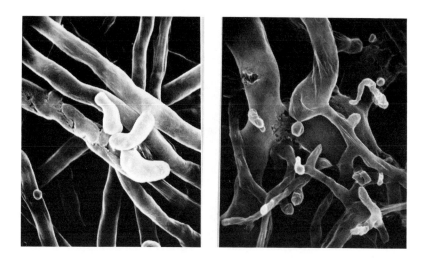

Fig. 2. Scanning electron micrographs showing the attachment of hyphal side branches of the mycoparasite, Pythium nunn, to the hyphae of P. aphanidermatum and Phytophthora parasitica. Note the host cell wall degradation. (X2500) (Lifshitz, Dupler, Elad, and Baker, source).

and Barkai-Golan et al.,[40] who reported binding of several lectins to hyphae of penicillia, aspergilli, and T. viride. Enzymatic degradation of the host cell walls is demonstrated in Fig. 2 with the mycoparasite, P. nunn.

Basically, there are two main types of interactions between Trichoderma and either R. solani or S. rolfsii. The mycoparasite may produce an appressorium-like body or it may coil around the host hyphae.[9,11] Shortly after the hyphal cells come together, the coiling hypha of Trichoderma constricts and partially digests the cell wall of S. rolfsii at the interaction site. Trichoderma then begins to penetrate the host cell. Penetration occurs at different locations on the host cell walls, even those which are thick and highly melanized.[15] Melanin is known for its ability to confer resistance to chemical digestion in fungi.[41] An example of these processes is given in Fig. 3. The digestion of the host cell walls may be due to the activity of extracellular enzymes, e.g., β-1,3-glucanases and chitinases, reported by Elad et al.,[31] with the same antagonistic fungus.

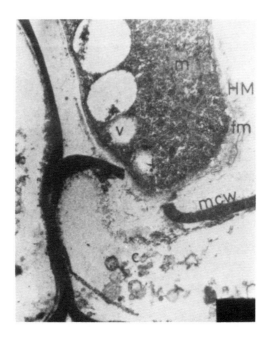

Fig. 3.   Penetration of Trichoderma into Rhizoctonia solani
hyphae observed in a transmission electron micrograph of a
thin section.   Note the constriction and the partial
digestion of the melanized cell wall (MCW) of the host.
Remaining cytoplasm (C) can be observed in R. solani cells
whereas vesicles (V), fibrillar materials (FM), and mitochondria are observed in Trichoderma cell (HM).   (X2000)
(Elad, Barak, Chet, and Henis, source).

      Relatively little information is available on the
sequence of events leading to the death of parasitized fungus
cell following cell wall penetration.   Shortly after the
host is penetrated by Trichoderma the penetrated hypha is
encapsulated by a thin electrolucent sheet.   The formation
of a sheath in the host cells in response to the invasion
of the parasite has also been reported for other mycoparasites.[21]   Mycoparasitism finally culminates in the loss of
the cytoplasmic contents of the host cells.   The remaining
cytoplasm, mainly surrounding the invading hyphae, exhibits
signs of disintegration (Fig. 3).   Disintegration of the

host cytoplasm proceeds rapidly after the wall is breached by the mycoparasite.

Antibiosis

Although many microorganisms produce antimicrobial substances in vitro, antibiosis as a mechanism of biocontrol of plant pathogens in the soil has not been demonstrated directly. However, it is reasonable to expect that anti-microbial substances enhance the capacity of their producers to colonize root surfaces and to compete for nutrients and space in the rhizosphere. Baker and Snyder[42] suggested that the effective range of these compounds is limited to an immediate area. The antibiotic gliotoxin, produced by T. viride, could only be detected in colonized seed coats or in bits of organic matter.[43] Howell and Stipanovic[44,45] provided evidence that different isolates of soilborne Pseudomonas fluorescens were antagonists to R. solani or to P. ultimum on cotton seedlings due to the production of the antibiotics, pyrrolnitrin [3-chloro-4-(2'-nitro-3'-chlorophenyl)-pyrrole] and pyoluteorin (4,5-dichloro-1 H-Pyrrole-2-yl-2,6-dihydroxyphenyl ketone), on cotton seeds. The bacteria were active when present in large numbers and had an available nutrient source.

Antibiotic substances also are produced by certain mycoparasites and, thus, may aid mycoparasitism. Culture filtrates of T. harzianum, and volatiles produced in cultures of the fungus inhibited the linear growth of P. aphanidermatum, S. rolfsii, and R. solani.[46,47] Contrary to the earlier reports, recent results obtained in Israel and by Windham and Baker in Colorado (personal communication) have shown that production by Trichoderma of a fungal inhibitory factor is isolate-dependent rather than species-dependent.[48,49] Among the Pythiaceae, P. mamillatum is known to inhibit the growth of a number of fungi in vitro including P. debaryanum and P. ultimum.[50] P. nunn which parasitizes Pythium sp. (N1) also is capable of inhibiting the formation of sporangia by Pythium sp. (N1) in soil through membranes. Elad, Lifshitz, and Baker (unpublished results) also isolated a water soluble, filterable factor from P. nunn which was capable of inhibiting the growth of Pythium sp. (N1) and R. solani in vitro, as well as the germination of sporangia of Pythium sp. in the soil. Morpho-logical changes which occurred in Pythium sp. (N1) following its contact with the factor included dissolution of germ

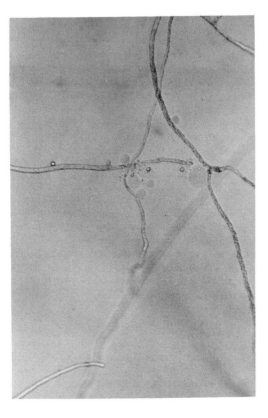

Fig. 4. Abnormal growth, swelling and bursting out of cytoplasm of Pythium sp. hyphae in the presence of a medium on which P. nunn has been cultured. (Elad, Lifshitz, and Baker, source).

tubes, rupturing of cytoplasmic membranes, swelling of hyphal tips, and the release of cytoplasm (Fig. 4). Similarly, Lifshitz, Windham, and Baker (personal communication) found that when Pythium sp. approached Trichoderma sp. in vitro, the protoplasm in the hyphal tip of Pythium sp. retreated and various signs of stress, such as distortion, vacuolation, and formation of septa, occurred in the mycelium.

The term "antibiosis" used by some[1,2,51] to describe the activity of the fluorescent siderophores of fluorescent

pseudomonads is not appropriate. These and other siderophores which inhibit or retard the growth of certain
microorganisms by competing for the essential iron will
be discussed in the following section.

Competition

    Microorganisms compete with each other for food and
for essential elements in the soil and around the rhizosphere.[52] Competition has, thus far, been documented for
nitrogen, carbon and iron.[53,54] A microbe may compete with
others by its ability to grow faster, or to take up and
assimilate the limited food quicker than other resident
microbes in a microsite. Competition may also be affected
by the ability of the competing microbes to produce
chelators of essential elements which bind the elements and
transport them into the cells. The best example of
chelator-mediated competition is provided by iron chelating
compounds (siderophores) which provide its producers with
an effective means to compete with other microbes for the
limited supply of iron. Siderophores are low molecular
weight iron transport molecules produced by most bacteria
and fungi in response to iron stress. There are two
classes of siderophores: the catechols, produced mostly
by bacteria, and the hydroxamates, synthesized generally
by fungi. Typically, three catechol or hydroxamate groups
occur in the same molecule and provide for very stable
hexadentate ligation of ferric ion with chelation constants
of about 10^{40} for the tricatechols[55] and less than 10^{30}
for the trihydroxamates.[56] Siderophores of the hydroxamate
class have a broad absorption peak at 440 nm whereas those
of the catechol class have a peak at 400 nm.[57,58]

    Since microorganisms which produce siderophores seem
to be capable of competing with other microorganisms for the
available iron in their environment, non-plant pathogenic
siderophore-producing microorganisms may potentially be used
as biological control agents to suppress the growth of plant
pathogens in the rhizosphere.

    Water-soluble, yellow-green fluorescent pigments
produced by many isolates of fluorescent pseudomonads in
certain media are the best examples of siderophores.[59-61]
Misaghi et al.[59,61] reported that a water-soluble fluorescent
siderophore produced by 156 isolates of fluorescent pseudomonads exhibited fungistatic activity against a number of

important plant pathogenic fungi including Geotrichum
candidum, R. solani, S. sclerotiorum, Phymatotrichum
omnivorum, Phytophthora megasperma, and P. aphanidermatum.
Kloepper et al.[60] showed that siderophores produced by
fluorescent pseudomonads inhibited the growth of Erwinia
carotovora and Escherichia coli. Sneh et al.[62] recently
reported that germination of chlamydospores of F. oxysporum
f. sp. cucumerinum was inhibited by fluorescent pseudomonads.
Addition of iron partially counteracted the inhibition of
germination. Moreover, this phenomenon was also observed
with other microelements, e.g., Zn^{2+}, Cu^{2+}, and Mn^{2+}. Elad
and Baker[63] have found that several cations competed with
Fe^{3+} on binding sites of the siderophore produced by pseudomonads. However, Misaghi et al.[59] observed no effect of
cations on the fungistatic activity of the siderophore
produced by pseudomonads. Pseudomonads reduced germination
of the fungus chlamydospores by 10-15% in soil at pH 5.0-
5.5, whereas addition of a partially purified siderophore
reduced germination by 41-43%. No inhibition occurred under
the same conditions upon addition of Fe^{3+} at a concentration of
2 mmol per gram of soil. Elad and Baker[63] simulated the
flow of root exudates in the rhizosphere (by addition of
1:4 concentrations of glucose-asparagine in pulses every 2
hr for 10 hr to the soil) and compared it to the application
of the total nutrients at one time. One application resulted
in up to 75% inhibition of chlamydospore germination; when
pulsed, inhibition was 15-35%. Similarly, the concentration
of root exudates is expected to have an effect on the activity of pseudomonads. Bacterial multiplication may be slow
in low nutrient rhizospheres of certain plants due to an
inability to compete with other microbes for the available
carbon and nitrogen sources. While the same number of cells
of fluorescent pseudomonads was produced in two diluted
media, siderophore production was reduced drastically in
more diluted medium as compared to non-diluted medium.

    Plant disease suppression by siderophore-producing
fluorescent pseudomonads has been demonstrated.[64-67] Take-
all disease of wheat was suppressed in both greenhouse and
field by treatment of wheat seeds with fluorescent pseudomonads.[66] The incidence of Fusarium wilt of carnations[68]
and of flax[64] was reduced when Pseudomonas putida was
introduced into the soil. The level of Fusarium wilt of
flax and radish also was dropped when the iron chelator,
EDDHA, was added to the soil.[64] Moreover, direct correlations were obtained between siderophore production by

pseudomonads in culture and their ability to induce suppressiveness in soil to Fusarium wilt of radish (r = 0.83), cucumber (r = 0.75), and peas (r = 0.87).[63] Inhibition of chlamydospore germination in soil of F. oxysporum f. sp. cucumerinum was greater in the presence of siderophore producing pseudomonads than in the presence of non-siderophore producing pseudomonads or those producing low levels of siderophores.[62]

The structure of a fluorescent siderophore produced by an isolate of Pseudomonas sp. has been elucidated. It is a linear hexapeptide linked through lysine to a quinoline derivative. A hydroxamate group and α-hydroxy acid and an O-dihydroxy aromatic group form the iron chelating portion of the molecule.[57]

Siderophores are also produced by fungi. F. oxysporum, a wilt-inducing pathogen, produces siderophores of the hydroxamate type under iron deficient conditions. This siderophore has a stability constant of $\log_{10} k = 29$, whereas the stability constant of the catechol hydroxamate siderophores of P. putida may be near $\log_{10} k = 40$.[64]

MECHANISMS OF PLANT GROWTH PROMOTION

Certain non-nitrogen fixing non-pathogenic bacteria are known to promote plant growth under certain conditions when they are applied to seeds or to roots. Among different bacterial genera tested, fluorescent pseudomonads seem to be the most effective.[1,60,67,69,70] Plant growth promotion also has been achieved by application of Trichoderma spp. to the soil (Keifeld, Baker, and Chet, personal communication and Baker et al.[71]).

The mechanism of plant growth enhancement mediated by non-nitrogen fixing growth promoting bacteria and fungi is not known. These microorganisms may produce growth regulators in the rhizosphere or induce plants to produce them in high quantities. They may solubilize certain unavailable minerals in the rhizosphere, or excrete substances that enhance seed germination, root growth, and production of lateral roots. Colonization of root surfaces by these microbes may increase the absorption rate of water and minerals from the soil by increasing the effective root surfaces and/or changing the electrochemical properties of

the roots. They may also change the chemical composition
of certain nutrients, e.g., transforming nitrate into
ammonium ion. Microbially-induced plant growth enhancement
may also be due to the ability of the growth promoting
microbes to protect plants from certain deleterious
rhizosphere-associated microorganisms.[3] Growth promoting
activity of Trichoderma observed in tobacco, tomato, and
radishes is due to the suppression of minor soil pathogens
such as Pythium spp. which cause the stunting of plants
and/or due to certain compounds which are excreted by the
fungus (Windham, Elad and Baker, unpublished). According
to Olsen and Misaghi,[67] the increased growth of guayule
seedlings following root inoculation with fluorescent
pseudomonads is probably not due to the release of nutrient
substances resulting from decomposition of the added bacterial cells. This is because heat-killed bacterial cells,
unlike live ones, failed to induce growth promotion in
guayule seedlings. Similar results were obtained by
Kloepper et al.[60]

CONCLUSIONS

     Much of the knowledge of the mechanisms of biological
control of plant pathogens and biological plant growth
promotion is derived from in vitro studies. Such findings
may not always be extrapolated into natural soil conditions
since all these mechanisms are influenced by biotic and
abiotic soil factors which vary from soil to soil. The
variability of biotic and abiotic factors in different soils
may explain why certain disease-protecting and plant growth-
promoting microorganisms are effective in the greenhouse
and not in the field. Among the soil factors influencing
the activity of the beneficial microbes, pH may be the most
important. The availability of iron to plant roots or
microorganisms is defined by the equilibrium reaction[72]

$$Fe(OH)_3 + 3H^+ \rightleftarrows Fe^{3+} + 3H_2O.$$

This equilibrium is modulated by the pH of the soil and,
therefore, Fe^{3+} ion will be limited in neutral or alkaline
soils. Since fluorescent pseudomonads produce fluorescent
siderophores only under low iron availability, little or no
siderophores are expected to be produced in a soil with low
pH and/or high iron levels. A biological control agent
which is dependent on siderophore production for its activity against a pathogen is probably ineffective in acid
soils.

The biological activity of the fluorescent siderophores also is subject to changes in pH value of the medium. Fungistatic activity of the fluorescent siderophores is increased drastically with increases in the pH value of the medium (Misaghi et al., unpublished results). The efficacy of the biological control agents which deploy cell wall degrading enzymes and/or antibiotics against their targets also is pH dependent, because the production of enzymes and antibiotics and/or their activities are often influenced by soil pH.

The production and the activity of siderophores may also be influenced by the soil levels of anions and cations. Elad and Baker[63] have found that cations, K^+, Na^+, NH_4^+, Ni^+, Ca^{2+}, Mn^{2+}, Zn^{2+}, Cu^{2+} and Mg^{2+} competed with Fe^{3+} on the binding site of the siderophore and/or inhibited the multiplication of the siderophore producing bacteria in soil and, therefore, delayed their development in log phase. However, Misaghi et al.[59] showed fungistatic activity of the fluorescent siderophores produced by fluorescent pseudomonad isolates was counteracted only by Fe^{3+} or Fe^{2+} but not by Al, Cr, Co, Ni, Pb, Mn, Zn, Cu, Mg, and Ca. Cations and anions in soil may also cause changes in the production and the activity of cell wall degrading enzymes and/or antibiotics produced by beneficial microorganisms.

Since the siderophores are secondary metabolites, a minimum concentration of nutrients is needed for their production.[56] The amount of fluorescent siderophores produced by fluorescent pseudomonads varies in different media and in the presence of different carbon and mineral sources.[73] Certain compounds in the soil or in the rhizosphere may serve as stimulators or suppressors of siderophore synthesis. Among the different carbon sources tested, glucose was found to be an effective suppressor of siderophore production by the fluorescent pseudomonads (Olsen and Misaghi, unpublished results).

The quality and the quantity of carbon and nitrogen sources and other chemicals in the environment may also modulate the efficacy of some beneficial microbes through their effects on their growth and on the production and activity of cell wall degrading enzymes, antibiotics as well as chelators. The chemical nature of root exudates may play an important role in this regard.[63]

The effect of certain soil parameters, discussed above, on the activities of biological control agents and of growth promoting bacteria is probably the main reason for their observed inconsistent performance from soil to soil. The performance consistency of beneficial biotic agents may be improved by construction of highly efficient microbes capable of combating plant pathogens with multiple weapons such as enzymes, antibiotics, chelators, and others. The efficacy of such super microbes to combat pathogens and to promote plant growth is less likely to be affected appreciably by changes in soil parameters.

The effectiveness of a microbially produced chelator to deprive a potential pathogen of the available vital elements depends to a large extent on the relative value of its stability constant compared to other chelators in the micro-environment. Because of their high stability constant, siderophores produced by fluorescent pseudomonads are expected to be effective chelators of iron. The ecological significance of hydroxamate siderophores in ferric iron mobilization in soil is apparent as compared with less specific chelators such as organic acids and phenolates which have stability constants of 10^{12}-10^{15}.[74]

REFERENCES

1. SCHROTH, M.N., J.G. HANCOCK. 1981. Selected topics in biological control. Annu. Rev. Microbiol. 35: 453-476
2. COOK, R.J., K.F. BAKER. 1983. The nature and practice of biological control of plant pathogens. Am. Phytopathol. Soc., St. Paul, Minnesota, 539 pp.
3. SCHROTH, M.N., J.G. HANCOCK. 1982. Disease suppressive soil and root colonizing bacteria. Science 216: 1376-1381.
4. BAKER, R. 1978. Inoculum potential. In Plant Disease, An Advanced Treatise. (J.G. Horsfall, E.B. Cowling, eds.), Vol. 2, Academic Press, New York, pp. 137-157.
5. PARK, D. 1960. Antagonism - the background of soil fungi. In The Ecology of Soil Fungi. (D. Parkinson, J.S. Waid, eds.), Liverpool Univ. Press, pp. 148-155.
6. HOMMA, Y., J.W. SITTON, R.J. COOK, K.M. OLD. 1979. Performation and destruction of pigmented hyphae and Gaeumanomycs graminis by vampyrellid amoebae from Pacific Northwest wheat field soils. Phytopathology 69: 1118-1122.

7.   ELAD, Y., I. CHET, J. KATAN. 1980. Trichoderma
     harzianum: a biocontrol agent effective against
     Sclerotium rolfsii and Rhizoctonia solani.
     Phytopathology 70: 119-121.
8.   ELAD, Y., Y. HADAR, E. HADAR, I. CHET, Y. HENIS. 1981.
     Biological control of Rhizoctonia solani by
     Trichoderma harzianum in carnation. Plant Dis.
     65: 675-677.
9.   ELAD, Y., I. CHET, P. BOYLE, Y. HENIS. 1983. The
     parasitism of Trichoderma spp. on plant pathogens-
     ultrastructural studies and detection by FITC
     lectins. Phytopathology 73: 85-88.
10. DENNIS, L., J. WEBSTER. 1971. Antagonistic properties
     of species groups of Trichoderma. III. Hyphal
     interaction. Trans. Br. Mycol. Soc. 57: 363-369.
11. CHET, I., G.E. HARMAN, R. BAKER. 1981. Trichoderma
     hamatum: its hyphal interactions with Rhizoctonia
     solani and Pythium spp. Microb. Ecol. 7: 29-38.
12. SEQUEIRA, L. 1980. Defense triggered by the invader:
     recognition and compatibility phenomena. In Plant
     Disease, An Advanced Treatise. (J.G. Horsfall,
     E.B. Cowling, eds.), Vol. 5, Academic Press, New
     York, pp. 179-200.
13. ELAD, Y., R. BARAK, I. CHET. 1983. The possible role
     of lectins in mycoparasitism. J. Bacteriol. 154:
     1431-1435.
14. HADAR, Y., Y. ELAD, Y. HENIS, I. CHET. 1982. Induction
     of macroscopic strands formation in Sclerotium rolfsii
     by Trichoderma harzianum. Israel J. Bot. 30: 156-164.
15. HARRIS, J.L., I.L. ROTH. 1980. Aerial strands of
     Phaeococcus exophialae. Can. J. Bot. 58: 562-567.
16. WATKINSON, S.C. 1979. Growth of rhizomorphs, mycelial
     strands, coremia and sclerotia. In Fungal Walls and
     Hyphal Growth. (J.H. Burnett, A.P.J. Trinci, eds.),
     Cambridge Univ. Press, Cambridge, pp. 93-113.
17. HENIS, Y., M. INBAR. 1968. Effect of Bacillus subtilis
     on growth and sclerotium formation by Rhizoctonia
     solani. Phytopathology 58: 933-938.
18. PENTLAND, G.O. 1967. Ethanol produced by Aureobasidium
     pullulans and its effect on the growth of Armillaria
     mellea. Can. J. Microbiol. 13: 1631-1635.
19. BUTTON, B., M. EL-KHOURI. 1978. Synnema and rhizomorph
     production in Sphaerostilbe repens under the influence
     of other fungi. Trans. Br. Mycol. Soc. 70: 131-136.
20. ELAD, Y., R. BARAK, I. CHET, Y. HENIS. 1983. Ultra-
     structural studies of interaction between Trichoderma

spp. and plant pathogenic fungi. Phytopathol. Z. 107: 168-175.

21. HOCH, H.C. 1978. Mycoparasitic relationships. IV. Stephanoma phaeospora parasitic in a species of Fusarium. Mycologia 70: 370-379.

22. MISAGHI, I.J. 1982. Physiology and Biochemistry of Plant-Pathogen Interactions. Plenum Press, New York, 287 pp.

23. BARTNICKI-GARCIA, S., E. LIPPMAN. 1973. In Handbook of Microbiology. (A.L. Laskin, H.L. Lechevalier, eds.), Vol. V, 2nd Edition, Chemical Rubber Co., Cleveland, Ohio, pp. 229-252.

24. MANOCHA, M.S. 1981. Host specificity and mechanism of resistance in a mycoparasitic system. Physiol. Plant Pathol. 18: 257-265.

25. TU, J.C. 1980. Gliocladium virens, a destructive mycoparasite of Sclerotinia sclerotiorum. Phytopathology 70: 670-674.

26. HOCH, H.C., M.S. FULLER, 1977. Mycoparasitic relationships. I. Morphological features of interactions between Pythium acanthicum and several fungal hosts. Arch. Microbiol. 11: 207-224.

27. HANSSLER, G.M., M. HERMANNS, H.J. REISENER. 1982. Elektronen-microskopische Beobachtungen der Interaktion zwischen Uredosporen von Puccinia graminis var. tritici und Verticillium lecanii. Phytopathol. Z. 103: 139-148.

28. LIFSHITZ, R., M.E. STANGHELLINI, R. BAKER. 1984. A new species of Pythium isolated from soil in Colorado. Mycotaxon. 20: 373-379.

29. LIFSHITZ, R., B. SNEH, R. BAKER. 1984. Soil suppressiveness to Pythium ultimum induced by antagonistic Pythium species. Phytopathology 74: 1054-1061.

30. ELAD, Y., R. LIFSHITZ, R. BAKER. 1984. Cell wall hydrolysis of host and non-host fungi during interaction with the mycoparasite, Pythium nunn. Phytopathology 74: 799 (abstract).

31. ELAD, Y., I. CHET, Y. HENIS. 1982. Degradation of plant pathogenic fungi by Trichoderma harzianum. Can. J. Microbiol. 28: 719-725.

32. BUMBIERIS, M. 1969. Effect of soil amendments on numbers of soil microorganisms and on the root rot Fusarium wilt complex of peas. Aust. J. Biol. Sci. 22: 1329-1336.

33. KLEINSCHUSTER S.J., R. BAKER. 1974. Lectin-detectable differences in carbohydrate-containing surface

moieties of macroconidia of Fusarium roseum
'Avenaceum' and Fusarium solani. Phytopathology
64: 394-399.
34. ELAD, Y., R. LIFSHITZ, M. DUPLER, R. BAKER. 1984.
Scanning electron and light microscopy of interaction between Pythium nunn and several soil fungi.
Proc. 49th Annu. Meeting, Mycological Society of
America. MSA Newsletter.
35. DEACON, J.W. 1976. Studies on Pythium oligandrum,
an aggressive parasite on other fungi. Trans. Br.
Mycol. Soc. 60: 383-391.
36. DRECHSLER, C. 1946. Several species of Pythium
peculiar in their asexual development. Phytopathology
36: 781-864.
37. BARNETT, H.L., F.I. BINDER. 1973. The fungal host-
parasite relationship. Annu. Rev. Phytopathol.
11: 273-292.
38. KRITZMAN, G., I. CHET, Y. HENIS, A. HUTTERMANN. 1978.
The use of the brightener "Calcofluor White M2R
New" in the study of fungal growth. Israel J. Bot.
27: 138-146.
39. MIRELMAN, D., E. GALUN, N. SHARON, R. LOTAN. 1975.
Inhibition of fungal growth by wheat germ agglutinin.
Nature 256: 414-416.
40. BARKAI-GOLAN, R., D. MIRELMAN, N. SHARON. 1978.
Studies on growth inhibition by lectins of penicillia
and aspergilli. Arch. Mikrobiol. 116: 19-24.
41. CHET, I., Y. HENIS. 1969. Effect of catechol and
disodium EDTA on melanin content of hyphal and
sclerotial walls of Sclerotium rolfsii Sacc. and the
role of melanin in the susceptibility of these walls
to β-(1,3)-glucanase and chitinase. Soil Biol.
Biochem. 1: 131-138.
42. BAKER, K.F., W.C. SNYDER, eds. 1965. Ecology of soilborne plant pathogens. Univ. of California Press,
Berkeley, 571 pp.
43. WRIGHT, J.M. 1956. The production of antibiotics in
soil. III. Production of gliotoxin in wheatstraw
buried in soil. Annu. Appl. Biol. 44: 461-466.
44. HOWELL, C.R., R.D. STIPANOVIC. 1979. Control of
Rhizoctonia solani on cotton seedlings with
Pseudomonas fluorescens and with an antibiotic
produced by the bacterium. Phytopathology 69:
480-482.
45. HOWELL, C.R., R.D. STIPANOVIC. 1980. Suppression of
Pythium ultimum - induced damping-off of cotton

seedlings by Pseudomonas fluorescense and its antibiotic, Pyoluteorin. Phytopathology 70: 712-715.

46. CHET, I., Y. ELAD. 1983. Biological and integrated
control of soilborne plant pathogens: mechanism and
application. Proceedings of Gottingen Symposium,
in press.

47. SIVAN, A., Y. ELAD, I. CHET. 1984. Biological control
effects of a new isolate of Trichoderma harzianum
on Pythium aphanidermatum. Phytopathology 74: 498-
501.

48. DENNIS, L., J. WEBSTER. 1971. Antagonistic properties
of species-groups of Trichoderma. I. Production of
non-volatile antibiotics. Trans. Br. Mycol. Soc.
57: 25-39.

49. DENNIS, L., J. WEBSTER. 1971. Antagonistic properties
of species-groups of Trichoderma. II. Production of
volatile antibiotics. Trans. Br. Mycol. Soc. 57:
41-48.

50. PARK, D. 1963. Evidence for a common fungal growth
regulator. Trans. Br. Mycol. Soc. 46: 541-548.

51. KLOEPPER, J.W., M.N. SCHROTH. 1981. Relationships of
in vitro antibiosis of plant growth-promoting
rhizobacteria to plant growth and the displacement
of root microflora. Phytopathology 71: 1020-1024.

52. BAKER, R. 1981. Ecology of the fungus Fusarium:
competition. In Fusarium: Disease, Biology and
Taxonomy. (P.E. Nelson, T.A. Toussoun, R.J. Cook,
eds.), Penn State Univ. Press, University Park and
London, pp. 245-249.

53. BAKER, R. 1968. Mechanisms of biological control of
soil-borne pathogens. Annu. Rev. Phytopathol. 6:
263-294.

54. BENSON, D.M., R. BAKER. 1970. Rhizosphere competition
in model soil systems. Phytopathology 60: '058-1061.

55. NEILANDS, J.B. 1973. Microbial iron transplant
compounds (siderophores). In Inorganic Biochemistry.
(I.G.L. Eickhorn, ed.), Elsevier, Amsterdam, pp. 167-
202.

56. EMERY, T. 1971. Role of ferrichrome as a ferric iono-
phore in Ustilago sphaerogena. Biochemistry 10:
1483-1488.

57. TEINTZE, M., M.B. HOSSAIN, C.L. BAINES, J. LEONG, D.
VAN DER HELM. 1980. Structure of ferric pseudo-
bactin, a siderophore from a plant growth promoting
Pseudomonas. Biochemistry 20: 6446-6457.

58. EMERY, T. 1969. Isolation, characterization and properties of fusarinine, a hydroxamic acid derivative of ornithine. Biochemistry 4: 1410-1417.

59. MISAGHI, I.J., L.J. STOWELL, R.G. GROGAN, L.C. SPEARMAN. 1982. Fungistatic activity of water-soluble fluorescent pigments of fluorescent pseudomonads. Phytopathology 72: 33-36.

60. KLOEPPER, J.W., J. LEONG, M. TEINTZE, M.N. SCHROTH. 1980. Enhanced plant growth by siderophores produced by plant growth-promoting rhizobacteria. Nature 286: 885-886.

61. MISAGHI, I.J., R.G. GROGAN, L.C. SPEARMAN, L.J. STOWELL. 1980. Antifungal activity of a fluorescent pigment produced by fluorescent pseudomonads. Proc. Am. Assn. Adv. Sci., Pacific Division, 61st Annu. Meet., p. 12.

62. SNEH, B., M. DUPLER, Y. ELAD, R. BAKER. 1984. Chlamydospore germination of Fusarium oxysporum f. sp. cucumerinum as affected by fluorescent and lytic bacteria from Fusarium suppressive soil. Phytopathology 74: 1115-1124.

63. ELAD, Y., R. BAKER. 1984. Influence of biomass, microelements and nutrient levels on activity of siderophore producing speudomonads in soil. Phytopathology 74: 806 (abstract).

64. SCHER, F.M., R. BAKER. 1983. Induction of suppressiveness in soil to Fusarium wilt pathogens with Pseudomonas putida and a synthetic iron chelate. Phytopathology 72: 1567-1573.

65. OLSEN, M.W., I.J. MISAGHI. 1982. Interaction among guayule, Verticillium dahliae, and non-pathogenic bacteria. Phytopathology 72: 935 (abstract).

66. WELLER, D.M., R.J. COOK. 1983. Suppression of take-all of wheat by seed treatments with fluorescent pseudomonads. Phytopathology 73: 463-469.

67. OLSEN, M.W., I.J. MISAGHI. 1984. Responses of guayule (Parthenium argantatum) seedlings to plant growth promoting pseudomonads. Plant Soil 77: 97-102.

68. SCHER, F.M., R. BAKER. 1980. Mechanism of biological control in Fusarium suppressive soil. Phytopathology 70: 412-417.

69. SUSLOW, T.V., M.N. SCHROTH. 1982. Rhizobacteria of sugar beets: effects of seed application and root colonization on yield. Phytopathology 72: 199-206.

70. BROWN, M.E. 1974. Seed and root bacterization. Annu. Rev. Phytopathol. 12: 181-197.

71. BAKER, R., Y. ELAD, I. CHET. 1984. The controlled
    experiment in the scientific method with special
    emphasis on biological control. Phytopathology 74:
    1019-1021.
72. LINDSEY, W.L. 1974. Role of chelation in micronutrient
    availability. In The Plant Root and Its Environment.
    (R.W. Carson, ed.), Univ. Press, Virginia,
    Charlottesville, pp. 507-524.
73. KING, J.V., J.J.R. CAMPBELL, B.A. EAGLES. 1948.
    Mineral requirements of fluorescein production by
    Pseudomonas. Can. J. Res. 266: 514-519.
74. POWELL, P.E., P.J. SZANISZALO, G.E. CLINE, C.P.P. REID.
    1982. Hydroxamate siderophores in the iron nutrition
    of plants. J. Plant Nutr. 5: 653-673.

Chapter Three

BIOCHEMICAL RESPONSES OF PLANTS TO FUNGAL ATTACK

LEROY L. CREASY

Department of Pomology
Cornell University
Ithaca, New York 14853

INTRODUCTION

This review will concentrate on developments in the literature since the publication of excellent reviews edited by Bailey and Mansfield[1] and by Wood[2] which should be consulted for a general overview. It is written with the preformed conclusion that active defense mechanisms are involved in fungi-plant interactions resulting in the resistance of some plants to some fungi.

Plants are continuously exposed to fungi. During first contact, fungi are often exposed to hostile meterological environments and most fungi succumb to the lack of satisfactory moisture or temperature conditions. When environmental conditions are favorable for the fungi, growth occurs and an initial interface is established between the plant and fungi. Successful interactions are those favoring fungal development while unsuccessful ones favor plant development. The definition of disease is involved therein. Fungi-plant interactions are represented by Figure 1. Initially significant in these interactions is the presence of often formidable physical barriers[3-5] composed of plant chemicals synthesized by previous biochemical activity.

47

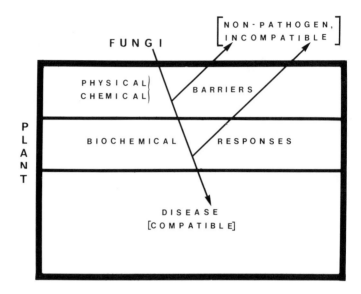

Fig. 1.   Fungi-plant interactions.

Within or beyond these barriers lie chemical barriers,
delineated here because their mode of action is chemical
rather than physical.   Chemical barriers are substances
normally present in the plant when it has not been exposed
to microorganisms (including but not limited to fungi).
Many such passive chemical barriers potentially involved
in fungi-plant interactions have been described.[4],[6-9]
Since most of the interest and certainly most of the
research on fungi-plant interactions has been accomplished
with food plants, we should remember that food plants have
been selected for the absence of many chemicals considered
unpalatable to animals, some of which may have been involved
in fungi-plant interactions in ancestral and in present-day
non-food plants.   One such plant component is tannin, which
if in contact with fungal extra-cellular enzymes would
surely cause inhibition.

     In a small number of cases, fungi successfully cross
the existing physical and chemical barriers and frequently
initiate sequences of events within the plant which will be
considered further.   These plant reactions can be termed
biochemical plant responses, and since they are usually
described because of their possible role in resisting rampant

fungal development, they can also be termed active plant
defenses. Why then should disease develop? Is it because
active plant defenses are not initiated? Is it due to the
suppression, or the ineffectiveness of the biochemical
response; e.g., fungal inactivation of the active defense
chemicals, or tolerance of the chemicals?

The concept of active plant defense against fungi
immediately evokes the concept of phytoalexins. Phytoalexins
are a carefully defined group of plant chemicals of low
molecular weight which are inhibitory to microorganisms and
whose accumulation in plants is initiated by interaction of
the plant with microorganisms. In this paper we will not
further define phytoalexins or attempt to evaluate whether
or not a particular chemical isolated from a plant is a
phytoalexin. Many plants produce several characteristic
phytoalexins. Phytoalexins in fact are part of a larger
group of chemicals which may be called active defense
chemicals. The latter term includes larger molecular
weight plant components but does not improve on the diffi-
culties involved in deciding whether a particular component
is an active defense chemical or not. As will be seen,
lignin is probably involved in the active biochemical
responses of plants resulting in resistance to fungi;
lignin is also a good example of a compound not fitting
the phytoalexin definition.

Disease will be considered as an interaction between a
plant and a fungus which is favoring the fungus. In
discussing diseases, only three categories of fungi will be
considered: (a) compatible-pathogens, those causing disease
in the non-manipulated plant; (b) incompatible-pathogens,
those belonging to species pathogenic on the plant which
for any reason do not cause disease in the specific non-
manipulated plant tested; and (c) non-pathogens, fungi which
do not cause disease on the non-manipulated plant species
studied. This simplification will disturb most pathologists
but it was done for the non-specialist who reads this
chapter.

Several key developments are leading to significant
progress in understanding fungi-plant interactions. The use
of elicitors, substances whose application to plants results
in activation of biochemical plant responses or disease
responses, has stimulated research greatly.[10,11] The exten-
sive use of plant cell cultures with or without elicitors

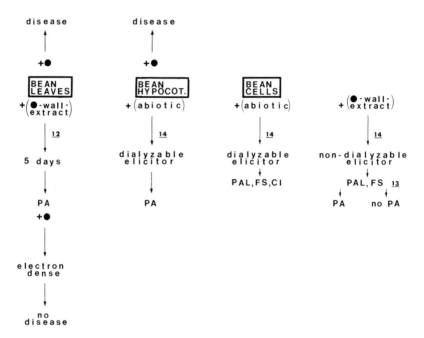

Fig. 2. Biochemical responses of bean species.
Abbreviations: phenylalanine ammonia-lyase, PAL; flavanone
synthase (= chalcone synthase), FS; chalcone isomerase, CI;
phytoalexin, PA. Numbers indicate appropriate references
to the literature. Solid circles represent compatible
pathogens.

provides needed cell homogeneity capable of simultaneous
biochemical stimulation. Histochemical techniques have
advanced steadily and are being exploited to good advantage
in understanding plant-fungi interactions. Immunological
techniques may expand the possibilities even further.

BIOCHEMICAL RESPONSES OF PLANTS TO FUNGI

Several examples of biochemical responses of plants to
fungal attack will be discussed. Different species of bean
share certain responses to fungal challenge (Fig. 2). Bean
leaves infiltrated with cell-wall glucans (elicitors)
prepared from a pathogen produced phytoalexin rapidly.

When leaves were inoculated 5 days later with spores of
the same pathogen, the plant was fully resistant. Fungal
development proceded normally after inoculation, with
spore germination and appressoria formation but then
stopped, accompanied by the appearance of an electron dense
material at the site of interaction.[12] This material,
possibly callose, may aid in the defense response by being
a site of phytoalexin accumulation.

Bean cell cultures treated with abiotic or biotic
elicitors synthesize phytoalexins and are ideal for the
measurement of enzymes involved in their biosynthesis.
In such cultures, phenylalanine ammonia-lyase (PAL) and
flavanone synthase (= chalcone synthase) activities
increased rapidly after treatment with elicitor. The
amount of enzyme activity was dependent on optimal concentrations of elicitor. Fractionation of the crude
pathogen-wall elicitor yielded several components, some of
which elicited PAL activity without phytoalexin synthesis.[13]
No fraction studied resulted in phytoalexin accumulation
without PAL elicitation however. The results suggest
specific elicitation of the several pathways required for
the synthesis of the phytoalexin.

Bean cell cultures were also used to show that an
abiotic elicitor added to bean cells resulted in dialyzable
elicitors of PAL activity. However, the factor responsible
for the biotic pathogen cell-wall elicitation response was
not dialyzable.[14] This system may provide a unique opportunity to examine elicitor activity. This is especially true
since the production of transportable or at least diffusible
elicitors is responsible for invoking phytoalexin synthesis
in plant cells sufficiently in advance of fungal development
to be useful in fungal inhibition.

Parsley cell cultures (Fig. 3) have proven useful in
many biochemical studies and appear equally useful in
investigations of fungi-plant interactions. Parsley cells
responded to a non-pathogen cell-wall glucan by producing
phenylalanine ammonia-lyase (PAL)[15-17] (mediated by
PAL-mRNA[15]), other enzymes of the phenylpropanoid pathway,[15-17] and dimethylallyldiphosphate:umbelliferone
dimethylallytransferase,[17] an enzyme specific for production
of the phytoalexins of this plant, furanocoumarins.[18] Cell
cultures respond to light by accumulating enzymes of the
phenylpropanoid pathway, enzymes of the flavonoid pathway and

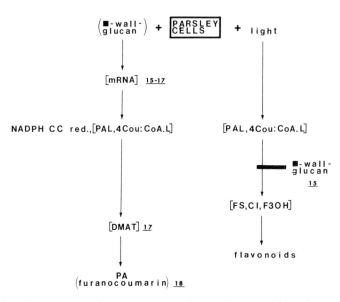

Fig. 3. Biochemical responses of parsley cell cultures.
Abbreviations: NADPH cytochrome c reductase, NADPH CC red.;
phenylalanine ammonia-lyase, PAL; 4-coumaryl:CoA ligase,
4Cou:CoA.L; chalcone isomerase, CI; flavanone synthase, FS;
flavonoid 3-hydroxylase, F3OH; dimethylallyldiphosphate:
umbelliferone dimethylallyl transferase, DMAT; phytoalexins,
PA. Solid squares represent non-pathogens.

flavonoids. On the other hand, the elicitor suppressed
accumulation of the flavonoid enzymes[15] and stimulated
accumulation of those enzymes directly involved in phytoalexin synthesis. This is a particularly exciting result
suggesting specific, selective pathway activation by
different inducers. The system should be useful in studying
not only elicitation but also the mechanisms which regulate
overlapping pathways.

The pea pod (Fig. 4) appears to be an ideal system for
examining the reaction of relatively intact plants to fungi.
Pea pods were shown to produce distinct groups of proteins
following elicitation,[19] inoculation with pathogens or
inoculation with non-pathogens.[20,21] Protein groups were
categorized by the translation in vitro of extracted mRNA.
One class of proteins resulted from inoculation with

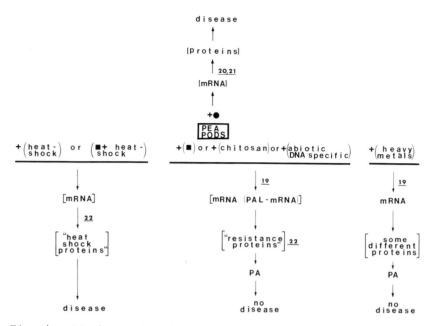

Fig. 4. Biochemical responses of pea pods. Abbreviations: phenylalanine ammonia-lyase, PAL; phytoalexins, PA. Solid circles and squares represent compatible pathogens and non-pathogens respectively.

pathogens; others called "resistance proteins" resulted from inoculation with non-pathogens, biotic or abiotic elicitors,[19] treatments leading to phytoalexin accumulation in the tissue. Based on the response to DNA-specific elicitors it was suggested that DNA participated directly in phytoalexin elicitation.[19] Heat shock frequently reduces or eliminates resistance mechanisms in the normal plant and heat shock of pea pods resulted in the production of yet another class of mRNA translatable into "heat shock proteins". Inoculation of these pods with non-pathogens resulted in disease. The production of heat-shock mRNA interfered with the production of resistance mRNA.[22] An additional group of proteins resulted from mRNA isolated from plants treated with heavy metal elicitors; these, although distinct from the resistance proteins, did result in phytoalexin synthesis and resistance.[19] This line of research may lead to an understanding of pathway elicitation

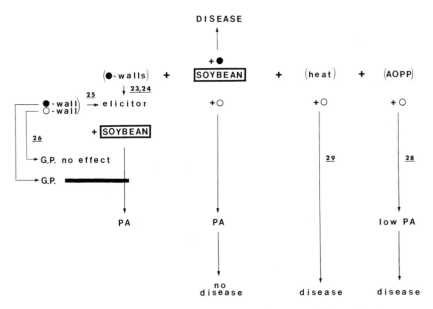

Fig. 5.   Biochemical responses of soybean.   Abbreviations:
L-α-aminooxy-β-phenylpropanoic acid, AOPP; glycoprotein,
G.P.; phytoalexin, PA.   Solid and open circles represent
compatible and incompatible pathogens respectively.

by fungi as the nature of the different protein groups is
determined.   PAL-mRNA is already identified in the reaction
leading to resistance.[19]   This is significant since PAL is
the first enzyme in the pathway leading to phytoalexin
synthesis.

Soybean (Fig. 5) is frequently used in studying
fungi-plant interactions.   Soybean tissues participate in
the production of an elicitor from pathogen cell-walls
which in turn results in the accumulation of phytoalexin
and resistance when applied to soybean.[23,24]   Elicitors
chemically prepared from compatible or incompatible
pathogens also stimulated phytoalexin synthesis.[25]   While
glycoproteins isolated from compatible pathogens inhibited
elicitor-activated, phytoalexin accumulation, glycoproteins
from non-compatible pathogens were not inhibitory.[26]
Suppression by a pathogen of an activated biochemical
reaction should be examined for its mechanism of action.

Other investigations have shown that the binding to soybean membranes of chemically modified elicitor fractions prepared from a non-pathogen was correlated with the ability of those fractions to elicite phytoalexin accumulation.[27]

Loss of resistance of soybean to incompatible fungi resulted from treatment of plants with the PAL inhibitor aminooxyphenylpropanoic acid (AOPP). This observation strengthens the involvement of phenylpropanoid metabolism in resistance.[28] The phytoalexin of soybean is synthesized by means of this pathway. Heat treatment interfered with the resistance response of soybean and the metabolism of the fungi involved.[29] The complex changes in disease development and phytoalexin concentration which followed different heat shock treatments are a good indication of the complexity of fungi-plant interactions.

The recent development of a specific immunological assay for detection and quantitation in vivo of the soybean phytoalexin[30] will permit more detailed study of the cellular interaction of compatible and incompatible pathogens.

Tomato plants (Fig. 6) contain large amounts of a preformed fungitoxic compound.[31] Tomato pathogens growing in the host plant appear resistant to this chemical but are inhibited by it when grown in free culture. Tomato pathogens are known to degrade this preformed, fungitoxic chemical and its role in compatible and incompatible interactions is unclear.[32]

Green tomato fruits respond to inoculation with an incompatible pathogen by an increased concentration of phenylpropanoids, increased activity of the phenylpropanoid enzyme PAL and increased deposition of lignin.[33] Lignin is considered a barrier to fungal development. The preformed fungitoxin, although not fully inhibitory to the fungus, may slow development enough during the initial interaction to permit synthesis of the lignin barrier.

Tomato leaves respond to inoculation with incompatible pathogens by hypersensitive cell death, callose deposition and phytoalexin production. In tomato, hypersensitive cell death is considered necessary for active resistance mechanisms. Compatible or incompatible pathogens grown in culture media produced non-race-specific elicitors of hypersensitive

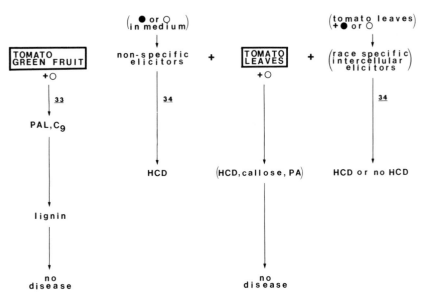

Fig. 6. Biochemical responses of tomato fruit or leaves.
Abbreviations: phenylalanine ammonia-lyase, PAL; phenylpropanoids, C_9; hypersensitive cell death, HCD; phytoalexins,
PA. Solid and open circles represent compatible and
incompatible pathogens respectively.

cell death in tomato leaves. But, inoculation of tomato
leaves with the same range of organisms resulted in race-
specific elicitors of hypersensitive cell death; these could
be collected from the intercellular fluids of the inoculated
tomato leaves. Incompatible pathogen-plant interactions
resulted in leaf diffusates causing hypersensitive cell
death while leaf diffusates collected from compatible
pathogen-plant interactions did not.[34] This is an example
of specific elicitors of defense response which are not only
pathogen-specific but are associated specifically with the
plant-fungi interaction.

Potato tubers (Fig. 7) have long been a popular
research material for studying the phytoalexin response.
Recently, investigators have challenged the association
between phytoalexins in potato and disease resistance.
Young, unstored tubers did not differ in their phytoalexin
synthesis although they demonstrated the expected disease

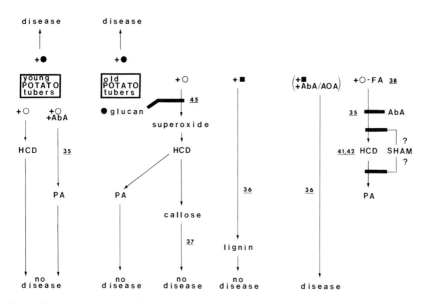

Fig. 7. Biochemical responses of potato tubers. Abbreviations: fatty acids, FA; α-aminooxyacetic acid, AOA; abscisic acid, AbA; O2⁻, superoxide anion; hypersensitive cell death, HCD; salicylhydroxamic acid, SHAM; phytoalexins, PA. Solid and open circles and solid squares represent compatible and incompatible pathogens and non-pathogens respectively.

reaction to compatible and incompatible pathogens.[35] The existence of additional resistance mechanisms was proposed. Abscisic acid (AbA), which elicited disease development when applied to stored tubers inoculated with incompatible pathogens,[36] elicited phytoalexin accumulation in interactions with incompatible pathogens in young tubers.[35]

Aminooxyacetic acid, an inhibitor of the phenylpropanoid enzyme PAL, which is not on the biosynthetic pathway to known potato phytoalexins also elicited disease when treated tubers were inoculated with a non-pathogen. It was shown that tubers showing resistant interactions had increased lignification at the plant-fungus interface which was not present in compatible interactions.[36] Encasement of fungal haustoria with callose material has also been observed in tuber interactions with incompatible pathogens.[37]

Two specific fatty acids identified in mycelia of an incompatible pathogen elicited hypersensitive cell death, phytoalexin accumulation,[38] and inhibited the continued accumulation of the constitutive steroid glycolalkaloids in potato tubers.[39] A pathogen cell-wall glucan enhanced the fatty acid elicitation of phytoalexins.[40] The elicitor-stimulated accumulation of phytoalexin was inhibited by AbA[35] and by salicylhydroxamic acid (SHAM).[41] However in the latter case it now appears impossible to determine if the inhibition by SHAM was due to its effect on lipoxygenase or on cyanide-resistant respiration.[42]

In potato tubers, hypersensitive cell death is thought to be prerequisite to the elicitation of phytoalexin synthesis and fungal restriction.[43] Studies with potato protoplasts indicated that the ability of compatible and incompatible pathogen hyphal-wall components to damage protoplasts was correlated with their pathogenicity. Inhibitors of hypersensitive cell death inhibited the response of protoplasts to the hyphal wall components.[44] Hypersensitive cell death has been correlated with the production of superoxide anion (O_2^-) by incompatible but not compatible pathogens.[45] Wall glucans from compatible pathogens inhibited superoxide anion generation.[46]

The study of phytoalexin production in potato cell cultures will contribute to clarification of the fungi-plant interactions of potato.[47]

A phenomenon which may be more common than is appreciated is exemplified by experiments with cucumber leaves (Fig. 8). Plant immunization to disease, the terminology carefully defined,[48] describes the protection of a plant from disease development as the result of prior exposure to an appropriate agent. When one leaf on a cucumber plant is inoculated with one of a limited list of microorganisms, the whole plant becomes resistant to later inoculation with pathogens. This immunization process results in quantitative systemic protection and was shown to be associated with increased phenylpropanoid production and the enhanced activity of some phenylpropanoid enzymes[49] and peroxidase.[50] No obvious, active resistance response is triggered until after initial fungal development where upon lignification proceeds rapidly in immunized plants but not in controls.[49,51] It was also shown that the mycelial walls of fungi can serve as a matrix for lignifi-

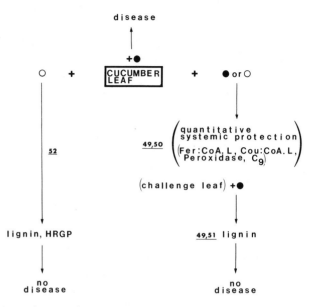

Fig. 8. Biochemical responses of cucumber leaves. Abbreviations: feruyl:CoA ligase, Fer:CoA.L; 4-coumaryl:CoA ligase, Cou:CoA.L; phenylpropanoids, C_9; hydroxyproline rich glycoproteins, HRGP. Solid and open circles represent compatible and incompatible pathogens respectively.

cation[51] in vitro. The concentration of cell wall glyco-protein rich in hydroxyproline increased during lignification of cucumber leaves in resistant cucumber cultivars but not in the disease interactions of susceptible cultivars.[52]

In oat leaves (Fig. 9), phytoalexins accumulated in collapsed cells after hypersensitive cell death and restricted fungal development; in compatible interactions, neither hypersensitive cell death nor phytoalexin accumu-lation occur.[53] Heat shock or treatment with aminooxyacetic acid inhibited phytoalexin accumulation and resulted in disease interactions with incompatible pathogens.[54] Protein labelling with ^{14}C-leucine during interactions revealed the synthesis of a group of proteins which were found in incompatible interactions and missing from compatible interactions.[55]

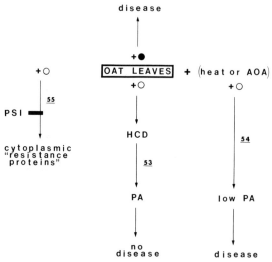

Fig. 9. Biochemical responses of oat leaves. Abbreviations: α-aminooxyacetic acid, AOA; protein synthesis
inhibitors, PSI; hypersensitive cell death, HCD;
phytoalexins, PA. Solid and open circles represent
compatible and incompatible pathogens respectively.

Lignification appears to be the biochemical response
in resistant interactions of fungi and wheat leaves (Fig.
10). This response is absent or suppressed in compatible
interactions.[56] The activity of phenylpropanoid enzymes
increase during reactions with non-pathogens.[57] Resistance
to pathogens could be induced in wheat leaves by treatment
with abiotic elicitors or pathogen germination fluids.
Subsequent inoculation with pathogens did not cause disease
and resulted in the production of fungitoxic agents in leaf
diffusates[58] although no phytoalexins are presently known
in wheat.

SYNOPSIS OF PLANT RESPONSES

Different types of active plant responses to fungi are
grouped in Table 1. Frequently it is difficult for the
non-specialist to distinguish between tissue death caused
by disease and hypersensitive cell death which precedes

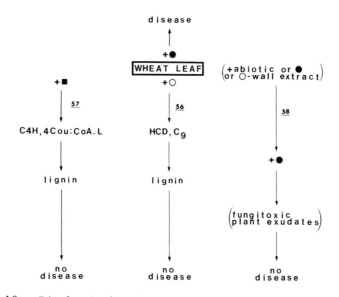

Fig. 10. Biochemical responses of wheat leaves.
Abbreviations: cinnamic acid 4-hydroxylase, C4H;
4-coumaryl:CoA ligase, 4Cou:CoA.L; hypersensitive cell
death, HCD; phenylpropanoids, C_9. Solid and open circles
and solid squares represent compatible and incompatible
pathogens and non-pathogens respectively.

Table 1. Categories of active defense mechanisms.

_____

Direct Phytoalexin Synthesis
    Pea pods
    Soybean cotyledons   (specific pathogen suppression)
    Bean leaves
    Bean cells
    Parsley cells

Hypersensitive Cell Death Followed by Phytoalexin
Accumulation
    Potato tubers   (specific pathogen suppression)
    Oat leaves   (specific pathogen suppression)
    Tomato leaves   (specific pathogen elicitation)

Lignification
    Cucumber leaves
    Wheat
    Potato tubers
    Green tomato fruit
_____

disease suppression (resistance) and is involved directly in
active biochemical responses.[59,60] In soybean for example,
hypersensitive cell death is not regularly reported during
resistant reactions. However, it has been pointed out that
the lack of macroscopic response was misleading since
limited hypersensitive cell death could be identified by
microscopic observation.[61]

It appears proper to consider that lignification is a
biochemical response resulting in resistance analogous to
the phytoalexin responses and appropriate to call both
categories active defense chemicals. The hypothesis is
that resistant responses involve synthesis of biochemicals
resulting in fungal limitation.

The existance of multiple mechanisms of resistance
should be anticipated. In support of the phytoalexin
theory, it should be noted that although resistance in the
absence of phytoalexin production was evident in the
literature, a lack of resistance was not observed when
phytoalexins were synthesized.

The sequence of response differs whether one considers
hypersensitive cell death as a prerequisite for the
resistant reaction or not. Most of the cases cited as
direct phytoalexin production were reactions with elicitors
while those involving phytoalexin production following
hypersensitive cell death were reactions with incompatible
pathogens.

A working hypothesis is that fungi contain recognition
compounds[62,63] (Ayers, this volume) which react only with
incompatible hosts. This reaction could result in directly
inducing the synthesis of active defense chemicals. It
could also result in specific host responses such as
hypersensitive cell death and the production of host-derived
elicitors of active defense chemicals. This hypothesis
provides a stimulus for chemists to identify a class of
host-derived elicitors which promote the synthesis of
phytoalexin without hypersensitive cell death. The existence
of such elicitors is suggested by those plant-derived
elicitors listed in direct phytoalexin stimulation.

According the above hypothesis, two major groups of
elicitors should be involved in interactions. One of these
would be fungal-produced elicitors, associated with identi-

fication of race specificity. These substances, in turn, initiate the production of the second group of plant-derived elicitors which would induce the production of active defense chemicals. The existence of two types of elicitors was suggested by the response of parsley cell culture to two elicitor fractions. One elicitor caused hypersensitive cell death and phytoalexin production while the other caused only phytoalexin synthesis.[17] According to this hypothesis, however, sucrose would be considered a plant-derived elicitor.[64]

Most research reviewed suggest the specificity of pathogens rather than hosts.

ACTIVATION OF DEFENSE CHEMICAL PRODUCTION

Plant responses to fungi resulting in the rapid synthesis of active defense chemicals are ultimately controlled by genetic information in each cell. Even when not under attack by fungi plants synthesize a plethora of chemical components and many of these are secondary plant chemicals of importance in plant metabolism and defense.[6,65]

A characteristic of successful, active, plant defense resulting from chemical synthesis is the rapid production and accumulation of defense compounds in concentrations which are effective. Such responses require rapid adjustments in enzymatic activities and the activation of dormant biosynthetic pathways. This rapid activation has in fact been helpful in studying the enzymology of biosynthetic pathways. Table 2 identifies the coordination between enzyme pathways which is required if the synthesis of the desirable end product is to proceed rapidly. The Table is, of course, highly oversimplified because within each category are modifying enzymatic reactions involving methylation, reduction, oxidation, acylation, prenylation, etc. Researchers should be able to demonstrate the activation of each major pathway during the synthesis of appropriate phytoalexins. For example, the synthesis of pterocarpanoid phytoalexins requires the coordinated activation (and regulation) of enzymes of the phenylpropanoid, acetate-malonate, flavonoid, isoflavonoid and pterocarpan pathways. Since some of these may already be activated for the synthesis of other chemicals, diversion of productivity or synthesis of enzymes in some

Table 2. Biochemical pathways activated or stimulated
during rapid biosynthesis of active defense chemicals.

	Class of Active Defense Chemical*				
Biochemical Pathway	L	S	F	I	P
Phenylpropanoid	X	X	X	X	X
Acetate-malonate		X	X	X	X
Stilbene (synthase)		X			
Flavonoid (synthase)			X	X	X
Isoflavonoid (aryl migration)				X	X
Pterocarpan (isoflavonoid 2'-hydroxylation)					X

*
L = lignin; S = stilbene; F = flavonoid; I = isoflavonoid;
P = pterocarpan.

cellular compartment specific for the synthesis of the
phytoalexin may be required.

A few examples of this kind of pathway activation were
mentioned earlier. Activities of PAL,[13,14,16,17,19,33]
cinnamic 4-hydroxylase[16] and 4-coumaryl:CoA ligase[15-17] are
increased under conditions leading to several categories of
active defense chemicals. PAL was found in several cases
to be synthesized de novo using the techniques of density
labelling[66] or by translation of plant mRNA in vitro.[15,19]
Additional enzymes involved in lignification such as
feruyl:CoA ligase[49] and peroxidase[50] increased in activity
in systemically protected cucumber. An enzyme catalyzing the
prenylation step in the biosynthesis of furanocoumarin-
phytoalexin increased in parsley cell cultures treated with
an elicitor.[17]

This system also provides an example of cross pathway
control. Parsley cell cultures normally synthesize flavonoids and exposure of cultures to light results in the
production of enzymes required for phenylpropanoids[67] and
flavonoids.[68] When illuminated cultures were treated with
the phytoalexin elicitor, those enzymes specific to
flavonoid biosynthesis were suppressed.[15] Activation of
phytoalexin synthesis therefore can involve not only

activation of required pathways but also the suppression of
redundant pathways. The factors controlling which pathway is
turned on or off during multiple stimuli and the order of
preference for different biosynthetic pathways has not been
investigated.

The synthesis of stilbene phytoalexins in grape was
shown to involve the coordinated production of phenylalanine
ammonia-lyase, cinnamic 4-hydroxylase and the specific
stilbene synthase.[69]

It is important to realize that pathway activation is
only the first step in the biosynthesis of active defense
chemicals. Individual enzymes of pathways are not stable
entities; they are continuously synthesized and degraded
(Fig. 11). A distinction should be made between (a) enzyme
synthesis, the rate at which the enzymatic protein is
produced, (b) enzyme turnover, the rate which an enzyme
loses activity through degradation, and (c) enzyme concentration, the amount of active enzyme available in the plant
at any time. The amount of active enzyme would be that
measured by extraction techniques if there are no inhibitors
of that enzyme in the tissue. The actual rate of conversion
of substrate to product in tissues may or may not be
described adequately by the extractable enzyme activity.
In addition, products known to be active defense chemicals
can be metabolized by fungi (degradation)[70,71] and also by
the plant (turnover).[72] Turnover is important in plant
metabolism because phytoalexins can be phytotoxic.[73] The
phytoalexin concentration measured at any one time provides
little information on rates of synthesis or degradation.

INDUCTION AND REGULATION OF PATHWAYS

Phenylpropanoid enzymes have received the most attention
in terms of pathway induction and rate regulation. It is
likely that other pathways will be found to be controlled by
the same or even more elegant mechanisms. Much of our
understanding has come from the study of plant cell cultures.
The subject of phenylpropanoid regulation has been well
reviewed by Hanson and Havir.[74] It is important to note
that the biochemical processes by which elicitors or other
inducers activate or induce a biosynthetic pathway are
unknown. It is usually agreed that PAL and cinnamic acid
4-hydroxylase (Fig. 12) are induced by light at the level

Fig. 11.   Instability of biosynthetic pathways and products.

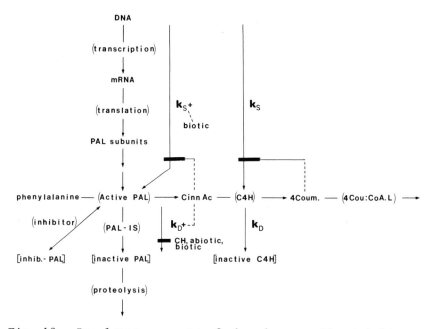

Fig. 12. Regulatory aspects of phenylpropanoid metabolism. Significant studies have been done only with PAL. Abbreviations: overall synthesis rate constant, k_S; biotic elicitor, biotic; phenylalanine ammonia-lyase, PAL; cinnamic acid, Cinn Ac; cinnamic acid 4-hydroxylase, C4H; 4-coumaric acid, 4Coum; 4-coumaric acid:CoA ligase, 4Cou:CoA.L; phenylalanine ammonia-lyase inactivating system, PAL-IS; overall degradation rate constant, k_d; cycloheximide, CH; abiotic elicitor, abiotic.

of transcription in tissues or in cell cultures which are not actively forming phenylpropanoid products. It is also agreed that the transient increase in phenylpropanoid enzyme activity is caused by an increase in the rate of enzyme synthesis (k_S) followed by an increase in the rate of enzyme degradation (k_d). Changes in the activity of PAL can be described by $dE/dt = k_S - k_d[E]$.[75] It appears likely that elicitor-stimulated PAL activity follows the transcription-translation sequence[19] although there is some uncertainty about the relative importance of transcription.

The rate of PAL synthesis was directly correlated with
the concentration of PAL-mRNA in all cases examined.
Results from several studies, however, suggest more direct
regulation of PAL-mRNA than is expected by means of
transcription. In bean suspension cultures, increases in
PAL activity caused by elicitors were highly dependent on
the elicitor concentration at the time of cell harvest and
not on the concentration during the lag phase.[13] PAL-mRNA
levels increased more rapidly in elicitor-treated parsley
cells than in illuminated cell cultures.[15] It therefore
seems possible that PAL synthesis may be regulated by a
mechanism which acts directly on the translation activity
of PAL-mRNA rather than the rate of mRNA production
following transcription.

Due to the transitory increases in PAL activity
frequently observed in cell cultures or plant tissues,
interest arose regarding the degradation phenomenon.
Originally suggested by Zucker in studies on potato tuber
tissue,[76] the concept has been applied to many experimental
systems. The details of turnover regulation are only now
beginning to be understood, although in the case of light-
stimulated phenylpropanoid synthesis in potato tuber slices,
it has been elegantly studied.

Stimulation by light of phenylpropanoid synthesis
involves transcription while the modulation of PAL activity
caused by pathway intermediates is mediated by a rapid
post-transcriptional mechanism.[77] It is agreed that PAL
induction is initiated by transcription and its rates of
synthesis regulated by mRNA.[79] Changes in pathway activity
brought about by inhibitors in potato,[79-81] elicitors in
bean cell cultures,[82] and wounding in pea epicotyls[83] were
interpreted as being due to effects on rates of enzyme
degradation.

Several systems have been described which degrade PAL
in vitro although their role in vivo is not established.
Sunflower leaves,[84,85] sweet potato roots,[86] and apple fruit
skin[87] have all been shown to contain a PAL inactivating
system in particulate fractions. Such inactivators may be
proteolytic enzymes which modify PAL in such a way that it
is then subjèct to general proteolytic hydrolysis. This
mechanism would be similar to the mechanism described for
tryptophan oxygenase in animal systems.[88] The combination
of change in the rate of enzyme synthesis together with

regulation of the rate of enzyme degradation would provide for rapid changes in enzyme availability necessary for pathways subject to sudden demands. Although inactivation systems appear to increase after PAL induction, they may also be regulated by changes in the susceptibility of their target enzyme since PAL is specifically protected from inactivation by its ligands. Pool size changes brought about by inhibitors could reduce or increase rates of degradation.

PAL and cinnamic 4-hydroxylase may also be subject to regulation by specific, soluble, proteinaceous inhibitors which bind reversibly to PAL. Such inhibitors were described in gherkin[89,90] and could provide for rapid changes in available, active enzyme. The inhibitor reported in barley[91,92] appears similar to the inactivators described above.

PAL frequently displays negative cooperativity kinetics, a property which would serve to dampen changes in reaction rate during changes in substrate availability. A temperature dependent, incompatible fungal interaction in sunflower leaves resulted in the production of a form of PAL which showed less negative cooperativity than the form encountered in compatible interactions.[93] Although observed, the significance of this finding is not obvious.

There is danger in relating changes in PAL activity to specific biochemical or disease reactions. This is illustrated by the report that PAL was induced in bean cell cultures by elicitor fractions which did not elicit phytoalexin synthesis.[13] In lettuce seedlings there was a close correlation between the effect of PAL inhibitors on radicle growth and PAL activity.[94] It should also be noted that there is a good correlation between hypersensitive cell death, lignification and PAL levels in lettuce induced to develop a physiological disorder.[95]

Considering our limited state of knowledge of the regulation of biosynthetic pathways, it is important to recognize the potential of multienzyme complexes and the channeling of biosynthetic intermediates.[96] A recent report suggests the association of both phenylpropanoid and flavonoid pathway enzymes with membranes.[97] The rate of flavonoid synthesis in some cell cultures is regulated by the concentration of active PAL, the first enzyme of a long

biosynthetic pathway.[98] The regulation of rates of synthesis
of active defense chemicals may be at biosynthetic points
which are not immediately obvious.

APPLICATIONS OF BIOCHEMICAL RESPONSES OF PLANTS

The large body of information available on the specific
biochemical responses of plants to fungi is unfortunately
dwarfed by the amount of information still needed. Although
hypotheses are useful in directing laboratory research,
concepts when established should lead to field applications.
The discussions found in recent publications imply several
potential applications. The pre-inoculation induction of
phytoalexins appears the least feasible due to phytoalexin
turnover, possible phytotoxicity or mammalian toxicity of
phytoalexins, etc. Some method of by-passing the pathogen
suppression of responses to resistance is attractive.
Conceptually, if pathogens cause non-response then treatment
of plants with substances which are hypothesized to react
with pathogens and prevent suppression could result in
disease resistance without preinduction of phytoalexins.
There is increasing interest in the possibility of spraying
non-fungitoxic materials which can provide protection from
fungi by eliciting (invoking) natural plant defense
mechanisms. Their mode of action is however still contro-
versial.[99] An attractive application of recent trends in
research appears likely in systemic protection reactions.[48]

CONCLUSIONS

Through understanding the induction and regulation of
biosynthetic pathways which produce active defense chemicals,
it should be possible to exploit the natural defense
mechanisms in plants, making them react more quickly and
in a less specific manner. The potential for application
of existing plant defenses lies within the reach of plant
pathologists and plant biochemists.

REFERENCES

1. BAILEY, J.A., J.W. MANSFIELD (eds.). 1982. Phytoalexins.
   John Wiley and Sons, New York, 334 pp.

2. WOOD, R.K.S. (ed.). 1982. Active Defense Mechanisms in Plants. Plenum Press, New York, 381 pp.
3. VANCE, C.P., T.K. KIRK, R.T. SHERWOOD. 1980. Lignification as a mechanism of disease resistance. Annu. Rev. Phytopathol. 18: 259-288.
4. FRIEND, J. 1981. Plant phenolics, lignification and plant disease. Prog. in Phytochem. 7: 197-261.
4. PEARCE, R.B., P.J. HOLLOWAY. 1984. Suberin in the sapwood of oak (Quercus robur L.): its composition from a compartmentalization barrier and its occurence in tyloses in undecayed wood. Physiol. Plant Pathol. 24: 71-81.
6. MAHADEVAN, A. 1982. Biochemical Aspects of Plant Disease Resistance. R.K. Jain, New Delhi, 397 pp.
7. COXON, D.T., S.K. OGUNDANA, C. DENNIS. 1982. Antifungal phenanthrenes in yam tubers. Phytochemistry 21: 1389-1392.
8. GIBSON, D.M., S. STACK. K. KRELL, J. HOUSE. 1982. A comparison of soybean agglutinin in cultivars resistant and susceptible to Phytophthora megasperma var. sojae (race 1). Plant Physiol. 70: 560-566.
9. CLINE, S., D. NEELY. 1984. Relationship between juvenile-leaf resistance to anthracnose and the presence of juglone glucoside in black walnut. Phytopathology 74: 185-188.
10. HOPPE, H.H. 1981. Auslosung der Phytoalexinanreicherung durch Infektionen order Elicitoren. Qualitas Plantarum 30: 289-302.
11. AHL, P. 1984. Lutte contre les microorganismes pathogènes des végétaux: les inducteurs de résistance. Phytopath. Z. 109: 45-64.
12. EBRAHIM-NESBAT, F., H.H. HOPPE, R. HEITEFUSS. 1982. Ultrastructural studies on the development of Uromyces phaseoli in bean leaves protected by elicitors of phytoalexin accumulation. Phytopath. Z. 103: 261-271.
13. DIXON, R.A., P.M. DEY, D.L. MURPHY, I.M. WHITEHEAD. 1981. Dose responses for Collectotrichum lindemuthianum elicitor-mediated enzyme induction in french bean cell suspension cultures. Planta 151: 272-280.
14. DIXON, R.A., P.M. DEY, M.A. LAWTON, C.J. LAMB. 1983. Phytoalexin induction in french bean. Intercellular transmission of elicitation in cell suspension cultures and hypocotyl sections of Phaseolus vulgaris. Plant Physiol. 71: 251-256.

15. HAHLBROCK, K., C.J. LAMB, C. PURWIN, J. EBEL, E. FAUTZ,
    E. SCHÄFER. 1981. Rapid response of suspension-
    cultured parsley cells to the elicitor from
    Phytophthora megasperma var. sojae. Plant Physiol.
    67: 768-773.
16. HAGMANN, M.-L., W. HELLER, H. GRISEBACH. 1983.
    Induction and characterization of a microsomal
    flavonoid 3'-hydroxylase from parsley cell cultures.
    Eur. J. Biochem. 134: 547-554.
17. TIETJEN, K.G., U. MATERN. 1983. Differential response
    of cultured parsley cells to elicitors from two
    non-pathogenic strains of fungi. 2. Effects on
    enzyme activities. Eur. J. Biochem. 131: 409-413.
18. TIETJEN, K.G., D. HUNKLER, U. MATERN. 1983.
    Differential response of cultured parsley cells to
    elicitors from two non-pathogenic strains of fungi.
    1. Identification of induced products as coumarin
    derivatives. Eur. J. Biochem. 131: 401-407.
19. LOSCHKE, D.C., L.A. HADWIGER, W. WAGONER. 1983.
    Comparison of mRNA populations coding for phenyl-
    alanine ammonia lyase and other peptides from pea
    tissue treated with biotic and abiotic phytoalexin
    inducers. Physiol. Plant Pathol. 23: 163-173.
20. HADWIGER, L.A., W. WAGONER. 1983. Electrophoretic
    patterns of pea and Fusarium solani proteins
    synthesized in vitro or in vivo which characterize
    the compatible and incompatible interactions.
    Physiol. Plant Pathol. 23: 153-162.
21. WAGONER, W., D.C. LOSCHKE, L.A. HADWIGER. 1982. Two-
    dimensional electrophoretic analysis of in vivo and
    in vitro synthesis of proteins in peas inoculated
    with compatible and incompatible Fusarium solani.
    Physiol. Plant Pathol. 20: 99-107.
22. HADWIGER, L.A., W. WAGONER. 1983. Effect of heat
    shock on the mRNA-directed disease resistance response
    of peas. Plant Physiol. 72: 553-556.
23. YOSHIKAWA, M., M. MATAMA, H. MASAGO. 1981. Release of
    a soluble phytoalexin elicitor from mycelial walls
    of Phytophthora megasperma var. sojae by soybean
    tissues. Plant Physiol. 67: 1032-1035.
24. KEEN, N.T., M. YOSHIKAWA. 1983. β-1,3-Endoglucanase
    from soybean releases elicitor-active carbohydrates
    from fungus cell walls. Plant Physiol. 71: 460-465.
25. NOTHNAGEL, E.A., M. McNEIL, P. ALBERSHEIM, A. DELL.
    1983. Host-pathogen interactions. XXII. A
    galacturonic acid oligosaccharide from plant cell

walls elicits phytoalexins. Plant Physiol. 71:
916-926.
26. ZIEGLER, E., R. PONTZEN. 1982. Specific inhibition
of glucan-elicited glyceollin accumulation in
soybeans by an extracellular mannan-glycoprotein
of Phytophthora megasperma f. sp. glycinea. Physiol.
Plant Pathol. 20: 321-331.
27. YOSHIKAWA, M., N.T. KEEN, M.-C. WANG. 1983. A
receptor on soybean membranes for a fungal elicitor
of phytoalexin accumulation. Plant Physiol. 73:
497-506.
28. MOESTA, P., H. GRISEBACH. 1982. L-2-Aminooxy-3-
phenylpropionic acid inhibits phytoalexin accumulation in soybean with concomitant loss of resistance
against Phytophthora megasperma f. sp. glycinea.
Physiol. Plant Pathol. 21: 65-70.
29. WARD, E.W.B., G. LAZAROVITS. 1982. Temperature-
induced changes in specificity in the interaction
of soybeans with Phytophthora megasperma f. sp.
glycinea. Phytopathology 72: 826-830.
30. MOESTA, P., M.G. HAHN, H. GRISEBACH. 1983. Development
of a radioimmunoassay for the soybean phytoalexin
glyceollin I. Plant Physiol. 73: 233-237.
31. DÉFAGO, G., H. KERN. 1983. Induction of Fusarium
solani mutants insensitive to tomatine, their
pathogenicity and aggressiveness to tomato fruits
and pea plants. Physiol. Plant Pathol. 22: 29-37.
32. SMITH, C.A., W.E. MacHARDY. 1982. The significance
of tomatine in the host response of susceptible and
resistant tomato isolines infected with two races
of Fusarium oxysporum f. sp. lycopersici.
Phytopathology 72: 415-419.
33. GLAZENER, J.A. 1982. Accumulation of phenolic
compounds in cells and formation of lignin-like
polymers in cell walls of young tomato fruits after
inoculation with Botrytis cinerea. Physiol. Plant
Pathol. 20: 11-25.
34. de WIT, P.J.G.M., G. SPIKMAN. 1982. Evidence for the
occurrence of race and cultivar-specific elicitors
of necrosis in intercellular fluids of compatible
interactions of Cladosporium fulvum and tomato.
Physiol. Plant Pathol. 21: 1-11.
35. BOSTOCK, R.M., E. NUCKLES, J.W.D.M. HENFLING, J.A. KUĆ.
1983. Effects of potato tuber age and storage on
sesquiterpenoid stress metabolite accumulation,
steroid glycoalkaloid accumulation, and response

74   LEROY L. CREASY

to abscisic and arachidonic acids. Phytopathology
73: 435-438.

36. HAMMERSCHMIDT, R. 1984. Rapid deposition of lignin
in potato tuber tissue as a response to fungi
non-pathogenic on potato. Physiol. Plant Pathol.
24: 33-42.

37. ALLEN, F.H.E., J. FRIEND. 1983. Resistance of potato
tubers to infection by Phytophthora infestans:
a structural study of haustorial encasement.
Physiol. Plant Pathol. 22: 285-292.

38. BOSTOCK, R.M., J.A. KUĆ, R.A. LAINE. 1981. Eicosapentaenoic and arachidonic acids from Phytophthora
infestans elicit fungitoxic sesquiterpenes in the
potato. Science 212: 67-69.

39. TJAMOS, E.C., J.A. KUĆ. 1982. Inhibition of steroid
glycoalkaloid accumulation by arachidonic and
eicosapentaenoic acids in potato. Science 217:
542-544.

40. MANIARA, G., R. LAINE, J.A. KUĆ. 1984. Oligosaccharides from Phytophthora infestans enhance the
elicitation of sesquiterpenoid stress metabolites
by arachidonic acid in potato. Physiol. Plant
Pathol. 24: 177-186.

41. ALVES, L.M., E.G. HEISLER, J.C. KISSINGER, J.M.
PATTERSON III, E.B. KALAN. 1979. Effects of
controlled atmospheres on production of sesquiter-
penoid stress metabolites by white potato tuber.
Possible involvement of cyanide-resistant respiration.
Plant Physiol. 63: 359-362.

42. STELZIG, D.A., R.D. ALLEN, S.K. BHATIA. 1983.
Inhibition of phytoalexin synthesis in arachidonic
acid-stressed potato tissue by inhibitors of
lipoxygenase and cyanide-resistant respiration.
Plant Physiol. 72: 746-749.

43. DOKE, N. 1982. A further study on the role of
hypersensitivity in resistance of potato cultivars
to infection by an incompatible race of Phytophthora
infestans. Physiol. Plant Pathol. 21: 85-95.

44. DOKE, N., N. FURUICHI. 1982. Response of protoplasts
to hyphal wall components in relationship to
resistance of potato to Phytophthora infestans.
Physiol. Plant Pathol. 21: 23-30.

45. DOKE, N. 1983. Involvement of superoxide anion
generation in the hypersensitive response of potato
tuber tissues to infection with an incompatible race

of Phytophthora infestans and to the hyphal wall
components. Physiol. Plant Pathol. 23: 345-357.
46. DOKE, N. 1983. Generation of superoxide anion by
potato tuber protoplasts during the hypersensitive
response to hyphal wall components of Phytophthora
infestans and specific inhibition of the reaction
by suppressors of hypersensitivity. Physiol. Plant
Pathol. 23: 359-367.
47. BRINDLE, P.A., P.J. KUHN, D.R. THRELFALL. 1983.
Accumulation of phytoalexins in potato-cell
suspension cultures. Phytochemistry 22: 2719-2721.
48. KUĆ, J. 1982. Induced immunity to plant disease.
BioScience 32: 854-860.
49. GRAND, C., M. ROSSIGNOL. 1982. Changes in the lignification process induced by localized infection
of muskmelons with Colletotrichum lagenarium.
Plant Sci. Lett. 28: 103-110.
50. HAMMERSCHMIDT, R., E.M. NUCKLES, J. KUĆ. 1982.
Association of enhanced peroxidase activity with
induced systemic resistance of cucumber to
Colletotrichum lagenarium. Physiol. Plant Pathol.
20: 73-82.
51. HAMMERSCHMIDT, R., J. KUĆ. 1982. Lignification as a
mechanism for induced systemic resistance in
cucumber. Physiol. Plant Pathol. 20: 61-71.
52. HAMMERSCHMIDT, R., D.T.A. LAMPORT, E.P. MULDOON. 1984
Cell wall hydroxyproline enhancement and lignin
deposition as an early event in the resistance of
cucumber to Cladosporium cucumerinum. Physiol.
Plant Pathol. 24: 43-47.
53. MAYAMA, S., T. TANI. 1982. Microspectrophotometric
analysis of the location of avenalumin accumulation
in oat leaves in response to fungal infection.
Physiol. Plant Pathol. 21: 141-149.
54. MAYAMA, S., S. HAYASHI, R. YAMAMOTO, T. TANI, T. UENO,
H. FUKAMI. 1982. Effects of elevated temperature
and α-aminooxyacetate on the accumulation of
avenalumins in oat leaves infected with Puccinia
coronata f. sp. avenae. Physiol. Plant Pathol. 20:
305-312.
55. YAMAMOTO, H., T. TANI. 1982. Two-dimensional analysis
of enhanced synthesis of proteins in oat leaves
responding to the crown rust infection. Physiol.
Plant Pathol. 21: 209-216.
56. BEARDMORE, J., J.P. RIDE, J.W. GRANGER. 1983.
Cellular lignification as a factor in the

hypersensitive resistance of wheat to stem rust.
Physiol. Plant Pathol. 22: 209-220.
57. MAULE, A.J., J.P. RIDE. 1983. Cinnamate 4-hydroxylase
and hydroxycinnamate:CoA ligase in wheat leaves
infected with Botrytis cinerea. Phytochemistry
22: 1113-1116.
58. CHAKRABORTY, D., A.K. SINHA. 1984. Similarity between
the chemically and biologically induced resistance
in wheat seedlings to Drechslera sorokiniana. Z.
Pflanzenkr. Pflanzenschutz. 91: 59-64.
59. HEALE, J.B., K.S. DODD, P.B. GAHAN. 1982. The induced
resistance response of carrot root slices to heat-
killed conidia and cell-free germination fluid of
Botrytis cinerea Pers. ex. Pers. 1. The possible
role of cell death. Annu. Botany 49: 847-857.
60. KUROSAKI, F., A. NISHI. 1984. Elicitation of
phytoalexin production in cultured carrot cells.
Physiol. Plant Pathol. 24: 169-176.
61. ÉRSEK T., M. HOLLIDAY, N.T. KEEN. 1982. Association
of hypersensitive host cell death and autofluo-
rescence with a gene for resistance to Peronospora
manshurica in soybean. Phytopathology 72: 628-631.
62. DIXON, R.A., C.J. LAMB. 1980. The specificity of plant
defenses. Nature 283: 135-136.
63. OUCHI, S. 1983. Induction of resistance or susceptibility. Annu. Rev. Phytopathol. 21: 289-315.
64. COOKSEY, C.J., P.J. GARRATT, J.S. DAHIYA, R.N. STRANGE.
1983. Sucrose: a constitutive elicitor of
phytoalexin synthesis. Science 220: 1398-1400.
65. SWAIN, T. 1977. Secondary compounds as protective
agents. Annu. Rev. Plant Physiol. 28: 479-501.
66. LAWTON, M.A., R.A. DIXON, C.J. LAMB. 1980. Elicitor
modulation of the turnover of L-phenylalanine
ammonia-lyase in French bean cell suspension
cultures. Biochim. Biophys. Acta 633: 162-175.
67. SCHRÖDER, J., B. BETZ, K. HAHLBROCK. 1976. Light-
induced enzyme synthesis in cell suspension
cultures of Petroselinum hortense. Demonstration
in a heterologous cell-free system of rapid
changes in the rate of phenylalanine ammonia-lyase
synthesis. Eur. J. Biochem. 67: 527-541.
68. SCHRÖDER, J., F. KREUZALER, E. SCHÄFER, K. HAHLBROCK.
1979. Concomitant induction of phenylalanine
ammonia-lyase and flavanone synthase mRNAs in
irradiated plant cells. J. Biol. Chem. 254: 57-65.

69. FRITZEMEIER, K.H., H. KINDL. 1981. Coordinate induction by UV light of stilbene synthase, phenylalanine ammonia-lyase and cinnamate 4-hydroxylase in leaves of Vitaceae. Planta 151: 48-52.

70. SMITH, D.A., J.M. HARRER, T.E. CLEVELAND. 1981. Simultaneous detoxification of phytoalexins by Fusarium solani f. sp. phaseoli. Phytopathology 71: 1212-1215.

71. ADIKARAM, N.K.B., A.E. BROWN, T.R. SWINBURNE. 1982. Phytoalexin involvement in the latent infection of Capsicum annuum L. fruit by Glomerella cingulata (Stonem.). Physiol. Plant Pathol. 21: 161-170.

72. FUJITA, M., K. ÔBA, I. URITANI. 1982. Properties of a mixed function oxygenase catalyzing ipomeamarone 15-hydroxylation in microsomes from cut-injured and Ceratocystis fimbriata-infected sweet potato root tissues. Plant Physiol. 70: 573-578.

73. BOYDSTON, R., J.D. PAXTON, D.E. KOEPPE. 1983. Glyceollin: a site-specific inhibitor of electron transport in isolated soybean mitochondria. Plant Physiol. 72: 151-153.

74. HANSON, K.R., E.A. HAVIR. 1981. Phenylalanine ammonia-lyase. In The Biochemistry of Plants. (E.E. Conn, ed.), Vol. 7, Academic Press, New York, pp. 577-625.

75. BETZ, B., E. SCHÄFER, K. HAHLBROCK. 1978. Light-induced phenylalanine ammonia-lyase in cell-suspension cultures of Petroselinum hortense. Quantitative comparison of rates of synthesis and degradation. Arch. Biochem. Biophys. 190: 126-135.

76. ZUCKER, M. 1968. Sequential induction of phenylalanine ammonia-lyase and a lyase-inactivating system in potato tuber disks. Plant Physiol. 43: 365-374.

77. LAMB, C.J. 1979. Regulation of enzyme levels in phenylpropanoid biosynthesis: characterization of the modulation by light and pathway intermediates. Arch. Biochem. Biophys. 192: 311-317.

78. FOURCROY, P. 1980. Properties of L-phenylalanine ammonia-lyase and turnover rate in etiolated and far-red illuminated seedlings of radish. Biochim. Biophys. Acta 613: 488-498.

79. LAMB, C.J. 1977. Phenylalanine ammonia-lyase and cinnamic acid 4-hydroxylase: characterisation of the concomitant changes in enzyme activities in illuminated potato tuber disks. Planta 135: 169-175.

80. LAMB, C.J., M.A. LAWTON, S.E. SHIELDS. 1981. Density
     labelling characterisation of the effects of cordy-
     cepin and cycloheximide on the turnover of
     phenylalanine ammonia-lyase. Biochim. Biophys.
     Acta 675: 1-8.
81. LAMB, C.J. 1982. Effects of competitive inhibitors
     of phenylalanine ammonia-lyase on the levels of
     phenylpropanoid enzymes in Solanum tuberosum.
     Plant, Cell and Environ. 5: 471-475.
82. DIXON, R.A., T. BROWNE, M. WARD. 1980. Modulation of
     L-phenylalanine ammonia-lyase by pathway interme-
     diates in cell suspension cultures of dwarf french
     bean (Phaseolus vulgaris L.). Planta 150: 279-285.
83. SHIELDS, S.E., V.P. WINGATE, C.J. LAMB. 1982. Dual
     control of phenylalanine ammonia-lyase production
     and removal by its product cinnamic acid. Eur. J.
     Biochem. 123: 389-395.
84. CREASY, L.L. 1976. Phenylalanine ammonia-lyase
     inactivating system in sunflower leaves. Phyto-
     chemistry 15: 673-675.
85. TAN, S.C. 1978. Effects of antibiotics and some
     other chemicals on phenylalanine ammonia-lyase and
     phenylalanine ammonia-lyase inactivating system in
     sunflower leaf. Sains Malaysiana 7: 171-181.
86. TANAKA, Y., I. URITANI. 1977. Synthesis and turnover
     of phenylalanine ammonia-lyase in root tissue of
     sweet potato injured by cutting. Eur. J. Biochem.
     73: 255-260.
87. TAN, S.C. 1979. Relationships and interactions
     between phenylalanine ammonia-lyase, phenylalanine
     ammonia-lyase inactivating system, and anthocyanin
     in apples. J. Amer. Soc. Hort. Sci. 104: 581-586.
88. SCHIMKE, R.T. 1969. On the roles of synthesis and
     degradation in regulation of enzyme levels in
     mammalian tissues. In Current Topics in Cellular
     Regulation. (B.L. Horecker, E.R. Stadtman, eds.),
     Academic Press, New York, pp. 77-124.
89. FRENCH, C.J., H. SMITH. 1975. An inactivator of
     phenylalanine ammonia-lyase from gherkin hypocotyls.
     Phytochemistry 14: 963-966.
90. BILLETT, E.E., W. WALLACE, H. SMITH. 1978. A
     specific and reversible macromolecular inhibitor
     of phenylalanine ammonia-lyase and cinnamic
     acid-4-hydroxylase in gherkins. Biochim. Biophys.
     Acta 524: 219-230.

91. PODSTOLSKI, A. 1981. Chloroplast-released inhibitor
     of phenylalanine ammonia-lyase from barley (Hordeum
     vulgare) seedlings. Physiol. Plantarum 52: 407-410.
92. PODSTOLSKI, A. 1983. Interaction between chloroplast
     and cytoplasmic factors in the inhibition of
     L-phenylalanine ammonia-lyase activity. Physiol.
     Plantarum 58: 107-113.
93. TENA, M., R.L. VALBUENA. 1983. Increase in phenyl-
     alanine ammonia-lyase activity caused by Plasmopara
     halstedii in sunflower seedlings resistant and
     susceptible to downy mildew. Phytopath. Z. 107:
     47-56.
94. KUDAKASSERIL, G.J., S.C. MINOCHA. 1984. Regulation of
     L-phenylalanine ammonia-lyase in germinating lettuce
     seeds: effect of substrate analogues and phenyl-
     propanoid compounds. Z. Pflanzenphysiol. 114: 163-
     171.
95. HYODO, H., H. KURODA, S.F. YANG. 1978. Induction
     of phenylalanine ammonia-lyase and increase in
     phenolics in lettuce leaves in relation to the
     development of russet spotting caused by ethylene.
     Plant Physiol. 62: 31-35.
96. STAFFORD, H.A. 1981. Compartmentation in natural
     product biosynthesis by multienzyme complexes. In
     E.E. Conn, ed., op. cit. Reference 74, pp. 117-137.
97. WAGNER, G.J., G. HRAZDINA. 1984. Endoplasmic
     reticulum as a site of phenylpropanoid and
     flavonoid metabolism in Hippeastrum. Plant Physiol.
     74: 901-906.
98. HAHLBROCK, K. 1981. Flavonoids. In E.E. Conn, ed.,
     op. cit. Reference 74, pp. 425-456.
99. BÖRNER, H., G. SCHATZ, H. GRISEBACH. 1983. Influence
     of the systemic fungicide metalaxyl on glyceollin
     accumulation in soybean infected with Phytophthora
     megasperma f. sp. glycinea. Physiol. Plant Pathol.
     23: 145-152.

Chapter Four

ALLELOPATHY - AN OVERVIEW

ELROY L. RICE

Department of Botany and Microbiology
University of Oklahoma
Norman, Oklahoma 73019

INTRODUCTION

Theophrastus[1] about 300 BC stated that chick pea
(Cicer arietinum) does not reinvigorate the ground as other
related plants (legumes) do but "exhausts" it instead. He
pointed out also that chick pea destroys weeds.

Pliny[2] in 1 AD reported that chick pea, barley (Hordeum
vulgare), fenugreek (Trigonella foenum-graecum), and bitter
vetch (Vicia ervilia) all "scorch up" cornland. Pliny
stated that walnut (apparently Juglans regia) causes head-
ache in man and injury to anything planted in the vicinity.
He gave numerous other examples of apparent chemical effects
of plants on other plants.

However in spite of these early observations concerning
apparent allelopathic effects, no solid scientific evidence
was obtained to support the suggestions until the present

century. The term <u>allelopathy</u> was coined by Molisch in 1937
to refer to biochemical interactions between all types of
plants, including microorganisms traditionally placed in the
plant kingdom.[3] In his discussion he indicated that he
meant the term to cover both inhibitory and stimulatory
biochemical interactions. Apparently most, if not all,
organic compounds which are inhibitory at a certain concentration are stimulatory to the same process at lower
concentrations. Undoubtedly, many important ecological
roles of allelopathy have been overlooked because of concern
with only the detrimental effects of the added chemicals.
Throughout this paper the term <u>allelopathy</u> is used according
to Molisch's broad definition.

A very important point concerning allelopathy is that
its effect depends on a chemical compound being added to
the environment. It is thus separated from <u>competition</u>
which involves ·the removal or reduction of some factor from
the environment that is required by some other plant or
microorganism sharing the habitat. Muller[4] suggested the
term <u>interference</u> to refer to the overall influence of one
plant (or microorganism) on another. Interference would
thus encompass both allelopathy and competition, and the
term interference is used here in this sense.

Evidence indicates that allelopathic compounds get out
of plants by volatilization, exudation from roots, leaching
from plants or residues by rain, or decomposition of
residues.[5]

SOME ROLES OF ALLELOPATHY IN NATURAL ECOSYSTEMS

Patterning of Vegetation

Probably most allelopathic effects of ecological
significance affect dispersion of plants and thus the
patterning. Generally this can only be demonstrated by
appropriate sampling procedures and statistical calcula-
tion, but in some instances the patterns are visually
apparent in the field.

Curtis and Cottam[6] observed the fairy-ring pattern of
the prairie sunflower, <u>Helianthus</u> <u>rigidus</u>, which is due to
a pronounced reduction in plant numbers, size, and inflo-
rescences in the center of the clone. They subsequently

demonstrated that the pattern was due to autotoxins produced
by decay of dead parts of the sunflower. Later, Wilson and
Rice[7] observed patterns of distribution of herbaceous
species around individuals of the common sunflower, Helianthus annuus, and demonstrated that the patterning was caused
primarily by allelopathic effects of the sunflower (Fig. 1).

Rasmussen and Rice[8] observed that a small grass,
Sporobolus pyramidatus, often expanded the size of its
stands in the University of Oklahoma Golf Course from a few
plants to large areas in a short time in spite of a heavy
stand of bermuda-grass, Cynodon dactylon, a more robust
plant (Fig. 2). They subsequently obtained strong evidence
that S. pyramidatus is able to spread rapidly into heavy
sods of bermuda-grass or buffalo-grass, Buchloe dactyloides,

Fig. 1. Zonation of species around Helianthus annuus in
field near Norman, Oklahoma. S, seedlings of H. annuus;
Bromus japonicus near H. annuus; Erigeron canadensis,
Rudbeckia serotina, and Haplopappus ciliatus in zone away
from H. annuus. Photographed by Dr. Roger Wilson.

Fig. 2.   Invasion of bermuda-grass stand by Sporobolus
pyramidatus (foreground).   Photographed by Dr. James
Rasmussen.

because it produced allelochemicals that are exuded from
living roots or diffuse from decaying roots or shoots and
inhibit seed germination and growth of these species.

Newman and Rovira[9] selected four grass and four forb
species from a permanent neutral British grassland for a
study of possible allelopathic interactions.   Water leachates of donor pots of each species were tested against each
of the eight species in receiver pots with extra nutrients
added on a regular schedule.   Leachates of all donor species
were significantly inhibitory compared with controls having
no plants in the donor pots.

Analysis of plants for N, P, and K showed that the
growth reductions were not due to nutrient deficiencies.
Four of the species were inhibited more by pot leachates of
their own species than by leachates from other species.
Three species gave the opposite response and one gave an
intermediate response to its own leachate.   Subsequent
field observations indicated that the most auto-inhibited

species are normally found as isolated individuals, or a few individuals in a group, not as pure stands. The three species which were allo-inhibited are all capable of dominating a permanent grassland. The authors concluded that allelopathy may play an important role in controlling species diversity in grasslands.

Prostrate knotweed, Polygonum aviculare, rapidly encroaches into bermuda-grass lawns and the bermuda-grass dies in patches of prostrate knotweed while bermuda-grass at the endges of the knotweed patches turns yellow. Soil minus litter was collected under a P. aviculare stand and under a bermuda-grass stand and was used to grow bermuda-grass.[10,11] Soil collected in March under knotweed markedly inhibited seed germination and seedling growth of bermuda-grass compared with soil from under bermuda-grass. Decaying roots and shoots of prostrate knotweed also reduced seed germination and seedling growth of bermuda-grass. Additionally, root exudates of knotweed reduced seedling growth of bermuda-grass. Eleven allelochemicals inhibitory to growth of bermuda-grass were isolated from soil under prostrate knotweed and none of these occurred in soil under bermuda-grass.[11,12] Four of the compounds were found to be phenolics and seven were long-chain fatty acids.

Japanese red pine, Pinus densiflora, forests are widespread in Japan and cover 60-70% of forest land in South Korea.[13] Vegetation under the trees is sparse despite the fact that the interior of these forests is one of the brightest among forests. Many other forests have dense undergrowths of herbs in spite of much lower light intensities. Various parts of red pine and the soil under it were found to contain chemicals toxic to many potential understory plants.[13,14] Thus, it was concluded that allelopathy probably plays an important role in retarding understory growth.

Striking patterns of vegetation occur in and around patches of Salvia leucophylla and Artemisia californica in the California chaparral.[15] Virtually no herbaceous species occur within the shrub stands, and a bare zone approximately 1-2 m wide occurs around the stands. Outside the bare zone is another zone 3-8 m wide which contains stunted plants of Bromus mollis, Erodium cicutarium, and Festuca megalura and this is surrounded by normal grassland (Fig. 3).[16] After numerous experiments, predation

Fig. 3. Salvia leucophylla producing differential composition in annual grassland: (1) to left of A, Salvia shrubs
1-2 m tall; (2) between A and B, zone 2 m wide bare of all
herbs except a few tiny seedlings of same age as the large
herbs to the right; (3) between B and C, zone of inhibited
grassland consisting of several grass species but lacking
Bromus rigidus and Avena fatua; (4) to right of C, uninhi-
bited grassland with large plants of numerous grass species
including Bromus rigidus and Avena fatua. (From Muller[16])

and competition for light, water, and minerals were eliminated as the basic cause of the pattern of inhibition.[16]
Several species of Salvia and Artemisia were found to
produce volatile compounds which inhibited growth of test
plants. Six monoterpenes were identified, and two of the
compounds, cineole and camphor, were consistently identified
in the air around Salvia leucophylla in the greenhouse and
in the field.[17]

    In the northern Negev Desert (Israel), annual plants
are more numerous on south-facing slopes dominated by
Zygophyllum dumosum than on the north-facing slopes dominated by Artemisia herba-alba (sage) in spite of the fact

that moisture conditions are more favorable on the north-facing slope.[18] Moreover, annuals are fewer in the vicinity of Artemisia than elsewhere. In the year following removal of sage the number of annuals increased but remained smaller than on the south-facing slope, suggesting a residual inhi-bitory effect. Tests demonstrated that shoots of sage released both volatile and water soluble substances which strongly inhibited germination of seeds of selected annual species which are rare in the populations of sage but common in adjacent associations. They were not affected by vola-tiles or water soluble materials from Zygophyllum dumosum. It was concluded, therefore, that allelopathy may be chiefly responsible for the absence or rarity of sensitive annual species in the neighborhood of A. herba-alba.

Plant Succession

As early as 1911, Cowles[19] emphasized the production of toxins by plants as a possible important factor in succes-sion. Relatively recent research has provided considerable evidence to support his hypothesis.

In the tall grass prairie region of Oklahoma and Kansas, there are four main successional stages when fields which are infertile are abandoned from cultivation: a pioneer weed stage which persists for only 2-3 years, an annual grass stage which persists for 9-13 years, a peren-nial bunch-grass stage which lasts for 30 years or longer after abandonment, and finally the climax prairie.[20] The evidence is strong that the pioneer weed stage disappears rapidly because the species are eliminated through strong allelopathic interactions.[5] Aristida oligantha, prairie threeawn, the dominant of the second stage, invades next apparently because it is not inhibited by the allelopathic chemicals produced by the pioneer weeds.[5] This species is able to grow well and reproduce in soil that is still too low in nitrogen and phosphorus to support species that invade later in succession.[21]

Aristida oligantha and several pioneer species produce allelochemicals which inhibit growth of Rhizobium and free-living nitrogen fixing organisms, and nodulation and leg-hemoglobin formation in legumes.[5] This indirect evidence suggested that the rate of biological nitrogen fixation was slowed in the first two stages of succession. Kapustka and Rice[22] measured nitrogen fixation rates in soils of the

pioneer weed stage, the annual grass stage, and the climax
prairie using the acetylene reduction technique. The rate
was about four times as high in the climax soil as in the
pioneer weed stage and about five times as high in the
climax as in the annual grass stage, thus substantiating the
indirect evidence. The slowing of the rate of nitrogen
fixation in the first two successional stages probably gives
A. oligantha a selective advantage in competition with
species having higher nitrogen requirements and causes it
to remain for a lengthy period.

Initially, it was thought that rapid nitrification was
necessary for the proper growth of the vegetation because
plant physiologists have always emphasized that nitrate is
the chief form of nitrogen available to plants. It was
surprising, therefore, to find that some climax species
inhibited nitrification more than any early invaders.[23]
Subsequently, Rice and Pancholy[24,25] found that concentrations of nitrate decreased from a high value in the first
successional stage to a low value in the climax prairie,
whereas concentrations of ammonium nitrogen increased from
a low value in the first stage to a high value in the
climax. Moreover, the counts of bacterial nitrifiers were
high in the first successional stage and decreased to a low
in the climax prairie. Thus, some factor or factors reduced
the populations of nitrifiers during succession resulting in
an apparent reduction in the rate of oxidation of ammonium
to nitrate. It was obvious from the general soil data[24]
that the low rates of nitrification in the climax plots were
not due to pH or textural differences. These facts, along
with the previous discovery that the climax species investigated were very inhibitory to nitrification, led to the
inference that the climax plants reduced the rate of nitrification. Such a reduction would help conserve nitrogen in
the ecosystem because the ammonium ion is positively charged
and is adsorbed by the negatively charged colloidal micelles
in the soil. The nitrate ion is negatively charged and is
repelled by the colloidal micelles. Thus, nitrate nitrogen
is readily leached below the depth of rooting or is washed
away into streams. Moreover, nitrate has to be reduced back
to ammonium before it can be used by the plant and this
requires energy. Thus, inhibition of nitrification conserves
both nitrogen and energy. There is evidence that tannins,
phenolic acids, flavonoids, and coumarin derivatives may be
important inhibitors of the process.[25,26]

During the completion of old-field succession to the
climax, the nitrogen concentration eventually increases to
the point where some later species can invade. This
apparently results in less inhibition of nitrogen fixation
and more inhibition of nitrification. Thus, the rate of
addition of nitrogen is increased and the rate of loss of
nitrogen is decreased. Eventually the concentration of
nitrogen is increased to the point where climax species
can invade.

Fields in the Piedmont of New Jersey abandoned after
spring plowing are invaded chiefly by ragweed, Ambrosia
artemisiifolia, and wild radish, Raphanus raphanistrum.[27]
They usually remain dominant for only one year after which
they are replaced by Aster pilosus as the dominant but with
a diverse group of secondary species. Ragweed and wild
radish failed to become re-established in plots cleared of
second stage perennial vegetation dominated by aster in
spite of the large number of seeds of these primary invaders
present in the soil. Analyses of the soil indicated that
the pattern of succession was not due to the mineral or
physical properties of the soil. Field soil from the second
stage inhibited germination and growth of ragweed and wild
radish whereas soil from the first stage did not. Root
exudate of ragweed and shoot extracts of ragweed and aster
inhibited germination and growth of early invaders. Thus,
it was concluded that the vegetational change from the first
to the second stage may be mediated by allelopathy.

SOME ROLES OF ALLELOPATHY IN FORESTRY

Walters and Gilmore[28] noted that height and growth of
sweetgum, Liquidambar styraciflua, was less in plots containing fescue, Festuca arundinacea, than in adjacent plots
without fescue. Chemical and physical soil factors did not
appear to explain the differences. Growth of sweetgum was
correlated with residual phosphorus and magnesium, but this
correlation was achieved across all experimental plots
without respect to the presence or absence of fescue. Seed-
ing of fescue into pots containing sweetgum seedlings
resulted in a reduction in dry weight increment of sweetgum
up to 95%. Elimination of competition through use of a
stairstep apparatus suggested that an allelopathic mechanism
was involved. Leachates from the rhizosphere of live fescue
and from dead fescue roots and leaves caused reductions in

Fig. 4.   A low-density upland cherry-maple orchard stand at
at Mill Creek, Ridgeway District, Allegheny National Forest.
Photo taken June 12, 1974 before much height growth had
occurred in the herbaceous plants.   Numerous stumps in
orchard stands are evidence of past productivity.   (From
Horsley[29])

dry weight increments of sweetgum up to 60%. Chemical
analysis of sweetgum seedlings from the stairstep experiment
suggested that fescue leachates decreased absorption of
phosphorus and nitrogen.

      Some forest areas in the Allegheny Plateau of north-
western Pennsylvania have failed to return to forest even
though 50 years have passed since clear-cutting (Fig. 4).[29]
Some of these contain a few scattered black cherry (Prunus
serotina) and red maple (Acer rubrum) trees and are called
orchard stands. Small black cherry seedlings grow slowly
and soon die in low density stands with a dense ground cover
of bracken fern (Pteridium aquilinum), wild oat grass
(Danthonia compressa), goldenrod (Solidago rugosa), and
flat-topped aster (Aster umbellatus). Browsing by animals,

microclimate, and competition were eliminated as the primary
causes of the inhibition of black cherry. Foliage extracts
of fern, goldenrod, and aster inhibited seed germination of
black cherry, and aster foliage extract inhibited both shoot
and root growth of black cherry seedlings growing on cotyledonary reserves. Foliage extracts of fern, grass, goldenrod
and aster also retarded shoot growth and dry weight accumulation of seedlings that had exhausted cotyledonary reserves.
In field tests, cherry seedlings did not survive and grow in
orchard soil even during the first year after removal of the
vegetation. They did survive during the second year after
removal, however, indicating that an appreciable time-span
was required for decomposition or leaching of the allelochemicals present in the soil.

Tubbs[30] found that sugar maple seedlings inhibited
growth of seedlings of yellow birch despite the apparent
absence of competition in nursery experiments. Root elongation of birch was retarded by exudates of actively growing
root tips of sugar maple. When seedlings of these species
were grown together in aerated nutrient solution, the number
of actively growing root tips of birch formed each day was
inversely correlated with the activity of the allelochemical
produced by maple as indicated by the retardation of elongation of yellow birch roots.

Alder species are often important in forests because of
the fixation of nitrogen by Frankia in nodules on their
roots. Jobidon and Thibault[31] observed growth depression
of alders near stands of balsam poplar, Populus balsamifera.
Water extracts of leaf litter and buds, and fresh leaf
leachates of balsam poplar inhibited seed germination and
radicle and hypocotyl growth of green alder (Alnus crispa
var. mollis) seedlings. There was marked inhibition of root
hair development and necrosis of the radicle meristems. The
average number of nodules on alder plants treated with
either of the three balsam extracts described above was only
51% of that of control plants.[32] Acetylene reduction
(nitrogen fixation) was decreased 62% by green alder plants
treated with the most concentrated bud and leaf litter
extracts.

Certain tree species such as Betula pendula and Picea
abies fail to develop in association with heather, Calluna
vulgaris.[33,34] This apparently results from the production
by heather of an allelochemical toxic to growth of

mycorrhizae on <u>Betula</u> and <u>Picea</u>. Fruticose soil lichens are
often allelopathic to the growth of mycorrhizae and forest
tree seedlings also.[35] Removal of reindeer moss (a lichen)
in field tests resulted in accelerated growth of pine and
spruce.

SOME ROLES OF ALLELOPATHY IN AGRICULTURE

Crop Plants Versus Crop Plants

Schreiner and his associates published several papers
shortly after 1900 which indicated that certain crop plants
produce compounds inhibitory to growth of the same and
other crop plants.[5] McCalla and Duley[36,37] reported the
allelopathic effects of decaying wheat residues in 1948-1949.
Many papers on allelopathic effects of crop plants have been
published in the past three decades, but only a few examples
will be given.

Clover soil sickness has been known in Europe since the
17th century,[38] and Tamura <u>et al.</u>[39,40] stated that it is
well known that red clover, <u>Trifolium pratense</u>, is allelopathic against itself. Nine inhibitory isoflavonoids or
related compounds were isolated from tops of red clover.[39,40]
All the identified isoflavonoids inhibited seed germination
of red clover at about 100 ppm and seedling growth at 10
ppm.[41] No isoflavonoids occurred in clover sick soil but
relatively high amounts of several allelopathic phenolic
compounds were found to be present.[41] These compounds were
later found to result from decomposition of the isoflavonoids present in red clover.

The unharvested parts of rice plants are generally
mixed with the soil because this has been thought to be
beneficial. However, it has been observed that productivity
of the second crop of rice in a paddy is less than that of
the first crop. Chou and Lin[42] found that aqueous extracts
of decomposing rice residues in soil retarded radicle growth
of rice seedlings and growth of rice plants. Maximum toxicity occurred in the first month of decomposition and
declined thereafter. Some toxicity persisted for 4 months
in the paddies. Five inhibitory phenolic acids were identified from decaying rice residues and several unidentified
allelochemicals were isolated.

In the southern part of Taiwan, a crop of rice is often followed immediately by a legume crop. Yields of soybeans have been increased by several hundred kilograms per hectare by burning the rice straw prior to planting the soybeans. Rice et al.[43] hypothesized that the decreased yields in unburned fields may result from an inhibition of nitrogen fixation by Rhizobium in the nodules of the soybean plants. The five phenolic acids identified by Chou and Lin[42] and sterile extracts of decaying rice straw in soil markedly inhibited growth of Rhizobium.[43] The phenolics also reduced nodule numbers and hemoglobin content of the nodules of two bean varieties. Moreover, extracts of decomposing rice straw in soil reduced nitrogen fixation (acetylene reduction) in Bush Black Seeded beans.

It has been observed for some time in Senegal in west Africa that the growth of sorghum is decreased markedly when it follows sorghum grown in sandy soils but not in soils high in montmorillonite.[44] Similar results occurred in the growth of sorghum seedlings when roots or tops of sorghum were added to sandy soils in laboratory experiments.[44] No inhibition resulted, however, when the residues were added to soil high in montmorillonite. Water extracts of roots or tops retarded growth of sorghum seedlings similarly. Inoculation with Trichoderma viride or an unknown species of Aspergillus eliminated the inhibitory effects of aqueous extracts of sorghum roots in a short time. Several weeks were required to detoxify non-sterile field soil after addition of root residues of sorghum. It was concluded that the microflora in the sandy soils of Senegal were not able to detoxify the soil fast enough to prevent inhibition of subsequent crops of sorghum.

Weeds Versus Crop Plants

Velvetleaf, Abutilon theophrasti, is a serious weed of several crops in the United States and Canada. Average yield reductions of soybeans under a variety of velvetleaf densities, placements, and duration of interference ranged from 14 to 41%.[45-47] Reductions in cotton yields ranged from 44 to 100%.[48,49] All the cited researchers attributed the reductions in crop yields to competition, although none performed experiments to determine whether allelopathy might be involved. Numerous other researchers have found velvetleaf to have marked allelopathic potential.[50-53] Water extracts of velvetleaf residues were slightly allelopathic

(5-24% inhibition) to radicle and coleoptile growth of corn
and to hypocotyl growth of soybeans.[52,53] Decaying residues
were highly allelopathic (50% or more inhibition) to height
growth and fresh weight increase of shoots of both corn and
soybeans in double pot experiments.[52,53]

Purple nutsedge, Cyperus rotundus, was listed by Holm[54]
as one of the ten worst weeds in the world. Interference by
this weed caused reductions in yields of various crops
ranging from 23 to 89%, even on the basis of average effects
of various weed densities.[5] It is noteworthy, therefore,
that numerous workers have found purple nutsedge to be
strongly allelopathic. Soil previously infested with this
weed for 9 to 12 weeks significantly reduced germination of
mustard, barley, and cotton seeds; and soil infested for
only 6 weeks significantly reduced germination of mustard
and cotton seeds.[55] Ethanol extracts of the previously
infested soil inhibited radicle growth of barley also.
Decomposing tubers of purple nutsedge reduced root and top
growth of barley,[56] sorghum,[57] and soybeans.[57]

Approximately 75 species of weeds have been reported to
have allelopathic potential but many of those have not been
tested against crop plants.[5] Moreover, the evidence for
allelopathic activity of many of the species is weak.

Allelopathic Effects of Crop Plants on Weeds

Both thin and dense field stands of Kentucky-31 fescue
were observed by Peters[58] to be relatively free of weeds.
Extracts of fescue, sand cultures, and split-root-system
experiments demonstrated that fescue produced toxic chemicals which exuded from the roots and inhibited growth of
wild mustard and birdsfoot trefoil.[58]

Three thousand accessions of the USDA collection of
oat, Avena, germplasm were screened for their ability to
exude scopoletin which is a compound known to have root
growth inhibiting properties.[59] Twenty-five accessions
exuded more blue-fluorescing material (characteristic of
scopoletin) from their roots than a standard oat cultivar
(Garry). Four accessions exuded up to three times as much
scopoletin as Garry oats. When one of these was grown in
sand culture for 16 days with a wild mustard, growth of the
mustard was significantly less than that obtained when the
weed was grown with Garry oats. Moreover, plants grown in

close association with the toxic accession exhibited severe chlorosis, stunting, and twisting indicative of chemical effects rather than competition. It appears feasible, therefore, to breed allelopathic genes into standard cultivars to aid in weed control.

Allelopathic crop plants have already been used experimentally in weed control. Leather[60] found one of thirteen genotypes of the cultivated sunflower tested to be very allelopathic to several weeds. In a 5-year field study with oats and sunflower grown in rotation, the weed density was significantly less than in control plots with oats only.

Putnam and DeFrank[61] tested residues of several fall and spring planted crops for weed control in Michigan. The plants were desiccated by the herbicides glyphosate or paraquat, or by freezing. Tecumseh wheat and Balboa rye residues reduced weed growth by up to 88%. Mulches of sorghum or sudangrass applied to apple orchards in early spring reduced weed biomass by 90% and 85%, respectively. In a 3-year series of field trials, sorghum residues reduced populations of common purslane by 70% and of smooth crabgrass by 98%.[62]

CHEMICAL NATURE OF ALLELOPATHIC COMPOUNDS

Allelopathic compounds consist of a wide variety of chemical types (Fig. 5).[5] It is evident from the diagram that these allelochemicals arise either through the acetate or shikimic acid pathway. These compounds range from very simple volatiles and aliphatic compounds to complex multiringed aromatic compounds. Only a few examples are mentioned below.

Several organic acids such as malic, citric, acetic, and tartaric acids in fruits are often concentrated enough to inhibit germination of seeds inside the fruits.[63] Unripe grains of corn and unripe seeds of peas will not germinate because of the presence of acetaldehyde.[63] Acetic and butyric acids were among the toxins produced during decomposition of rye residues,[64] and salts of acetic, propionic, and butyric acids were the chief phytotoxins produced in decaying wheat straw.[65]

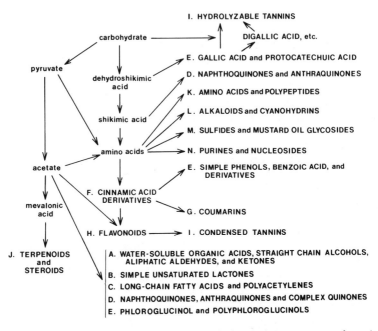

Fig. 5. Chemical categories of allelopathic compounds with probable major biosynthetic pathways leading to their production. (From Rice[5])

A simple lactone, parasorbic acid, from the fruit of mountain ash, inhibits seed germination and also has anti-bacterial action.[63] Another such compound, patulin, is produced by several fungi, including Penicillium urticae, which produced large amounts of the substance when growing on wheat straw.[66]

Long chain fatty acids have long been reported to be important allelochemicals produced by algae.[67] These compounds were recently reported to be potent toxins in decaying residues of a higher plant, Polygonum aviculare.[12] Polyacetylenes are apparently derived from long chain fatty acids,[68] and evidence is increasing that they are important and powerful allelopathic compounds.[69,70] α-Terthienyl produced by roots of marigold, Tagetes erecta, caused 50% mortality in seedlings of four test species in concentrations from 0.15 to 1.93 ppm.[70] Many of these compounds are antimicrobial also.[5] Certain polyacetylenes or derivatives offer excellent possibilities as herbicides.

Juglone is the only quinone identified as an allelopathic compound from higher plants.[5] It is produced by
walnut trees and is a potent inhibitor. Numerous antibiotics produced by microorganisms are quinones, including
the tetracycline antibiotics such as aureomycin.[71]

Simple phenols, phenolic acids derived from benzoic
acid, and those derived from cinnamic acid have been the
most commonly identified allelopathic compounds produced
by higher plants.[5] The most common allelopathic compounds
identified in soil under allelopathic plants are
p-hydroxybenzoic, vanillic, p-coumaric, and ferulic
acids.[5] This may be due to the ease with which these
compounds can be identified, and even more important allelochemicals may have been overlooked.

Coumarins are lactones of o-hydroxycinnamic acid in
which side chains often are isoprenoids.[68] Coumarin,
esculin, and psoralen (a furanocoumarin) all strongly
inhibit seed germination. Such inhibitors are produced by
a variety of legumes and cereal grains.

Flavonoids are widespread in higher plants and a
few have been implicated in allelopathy.[5] The flavonoid
phlorizin, present in apple roots, is toxic to young apple
trees and often causes difficulty in replanting old apple
orchards. Numerous flavonoids and their glycosides are
produced by species from the tall grass prairie and post
oak-blackjack oak forest and are inhibitory to nitrifying
bacteria and to send germination.[26]

Several hydrolyzable and condensed tannins have been
implicated in allelopathy.[5] They have been identified as
growth and germination inhibitors in dry fruits,[72] as growth
retarders of nitrogen fixing and nitrifying bacteria in
several plants, and as reducers of seedling growth.[5]

Higher plants produce a great variety of terpenoids[68]
but only a few have been implicated in allelopathy. The
monoterpenoids are the major components of essential oils of
higher plants[68] and they are the predominant terpenoid inhibitors that have been identified.[5] Some of the roles of
these compounds in the patterning around various shrubs in
the California Chaparral have been discussed and these also
contribute to allelopathy in eucalyptus and sagebrush.[5] Many
fungi[73] and algae[74] also produce terpenoid allelochemicals.

There are only a few instances in which amino acids
have been implicated in allelopathy and in most cases the
specific amino acids have not been identified. Rhizobitoxine
is produced by certain strains of Rhizobium japonicum and is
a non-protein amino acid.[75] Several of the phytotoxins
produced by pathogenic microorganisms are polypeptides and
related glycopeptides.[73]

Many alkaloids have been implicated in plant-animal
chemical interactions but few have been associated with
allelopathy.[76] Several alkaloids were demonstrated by
Evenari[63] to be strong inhibitors of seed germination.
Little recent work has been done on alkaloids except for
caffeine.[68] α-Picolinic acid is a microbial alkaloid with
toxic action on plants.[73] One of the more active synthetic
herbicides on the market, picloram (Dow's Tordon), is a
chlorinated picolinic acid derivative.

Cyanohydrins have been implicated in allelopathy in
several instances. Dhurrin occurs in grain sorghum seedlings and the seedlings contain enzymes that hydrolyze
dhurrin to glucose, HCN (hydrogen cyanide) and p-hydroxybenzaldehyde.[77] The situation is similar in Johnsongrass,
Sorghum halepense, a very allelopathic weed.[78] Both HCN
and p-hydroxybenzaldehyde are potent allelochemicals. HCN
and benzaldehyde are produced by the hydrolysis of amygdalin
which is present in peach root residues.[79] HCN and benzaldehyde are inhibitory to growth of peach seedlings and
apparently cause the peach replant problem in old peach
orchards.[79]

Mustard oils, such as allyl-isothiocyanate, are products of the hydrolysis of mustard oil glycosides.[68] Mustard
oils are produced by all organs of plants belonging to the
Cruciferae (mustard family),[63] and are strong inhibitors of
seed germination and microbial growth.

Many antibiotics produced by various microorganisms are
nucleosides.[5] Among these are nebularine, cordycepin, and
nucleocidin. The only known purines in higher plants shown
to be involved in allelopathy are caffeine, theophylline,
paraxanthine, and theobromine from the coffee bush.[80]
Evenari[63] included caffeine as an alkaloid and stated that
it is one of the most potent alkaloids in the inhibition of
seeds.

There are many cases in which significant allelopathic
effects are known but in which the allelochemicals have not
been identified. Techniques are available now to speed up
identification of allelopathic compounds and this is very
important in determining methods of egress of the chemicals
from organisms, amounts present in the environment, amounts
absorbed by affected organisms, and methods and rate of
decomposition.

CONCLUSION

Interest in allelopathic research has accelerated
rapidly in the past decade and is high worldwide. Research
has been particularly active concerning the roles of
allelopathy in plant succession, the patterning of vegetation, forestry, and agriculture. Increasing knowledge of
allelopathy is aiding greatly in our understanding of many
ecological phenomena. Increasing knowledge of the conditions
under which certain crop residues cause allelopathic effects
on subsequent crops should enable us soon to guard against
such effects. We are on the threshold of breeding crop
plants that will inhibit growth of the chief weeds in a given
area through allelopathic action and thus decrease the need
for synthetic weed killers. We are already able to use residues of allelopathic crop plants and weeds to control weed
growth in some crops and in orchards. Our understanding of
allelopathic interactions between various plant species has
been used advantageously in reforestation, and future developments are encouraging.

Evidence is mounting that inhibition of nitrification
increases as succession progresses toward the climax vegetation, at least in many vegetation types. This leads to a
decrease in the loss of nitrogen. Addition of nitrification
inhibitors to arable lands has proved to be valuable in
preventing loss of nitrogen and in increasing crop yields.

The horizons for future research in allelopathy are
unlimited and applications of the knowledge gained will be
phenomenal.

REFERENCES

1. THEOPHRASTUS. ca 300 BC. Enquiry into Plants and
    Minor Works on Odours and Weather Signs. (English
    translation by A. Hort.) 2 Vols. W. Heinemann,
    London, pp. 1916.
2. PLINIUS SECUNDUS, C. 1 AD. Natural History. (English
    translation by H. Rackam, W.H.S. Jones, and D.E.
    Eichholz.) 10 Vols. Harvard University Press,
    Cambridge, Massachusetts, pp. 1938-1963.
3. MOLISCH, H. 1937. Der Einfluss einer Pflanze auf die
    andere-Allelopathie. Gustav Fischer, Jena.
4. MULLER, C.H. 1969. Allelopathy as a factor in
    ecological process. Vegetatio 18: 348-357.
5. RICE, E.L. 1984. Allelopathy, Second Edition.
    Academic Press, Orlando, Florida, 400 pp.
6. CURTIS, J.T., G. COTTAM. 1950. Antibiotic and auto-
    toxic effects in prairie sunflower. Bull. Torrey
    Bot. Club 77: 187-191.
7. WILSON, R.E., E.L. RICE. 1968. Allelopathy as
    expressed by Helianthus annuus and its role in old-
    field succession. Bull. Torrey Bot. Club 95:
    432-448.
8. RASMUSSEN, J.A., E.L. RICE. 1971. Allelopathic
    effects of Sporobolus pyramidatus on vegetational
    patterning. Am. Midl. Natur. 86: 309-326.
9. NEWMAN, E.I., A.D. ROVIRA. 1975. Allelopathy among
    some British grassland species. J. Ecol. 63:
    727-737.
10. ALSAADAWI, I.S., E.L. RICE. 1982. Allelopathic
    effects of Polygonum aviculare L. I. Vegetational
    patterning. J. Chem. Ecol. 8: 993-1009.
11. ALSAADAWI, I.S., E.L. RICE. 1982. Allelopathic
    effects of Polygonum aviculare L. II. Isolation,
    characterization and biological activities of
    phytotoxins. J. Chem. Ecol. 8: 1011-1023.
12. ALSAADAWI, I.S., E.L. RICE, T.K.B. KARNS. 1983.
    Allelopathic effects of Polygonum aviculare L.
    III. Isolation, characterization, and biological
    activities of phytotoxins other than phenols. J.
    Chem. Ecol. 9: 761-774.
13. LEE, I.K., M. MONSI. 1963. Ecological studies on
    Pinus densiflora forest 1. Effects of plant
    substances on the floristic composition of the
    undergrowth. Bot. Mag. (Tokyo) 76: 400-413.

14. KIL, B.S. 1981. Allelopathic effect of Pinus densiflora on the floristic composition of under-growth in pine forests. Doctoral Dissertation, Department of Biology, Chung-Ang University, Iri, Korea.

15. MULLER, C.H., W.H. MULLER, B.L. HAINES. 1964. Volatile growth inhibitors produced by shrubs. Science 143: 471-473.

16. MULLER, C.H. 1966. The role of chemical inhibition (allelopathy) in vegetational composition. Bull. Torrey Bot. Club 93: 332-351.

17. MULLER, C.H. 1965. Inhibitory terpenes volatilized from Salvia shrubs. Bull. Torrey Bot. Club 92: 38-45.

18. FRIEDMAN, J., G. ORSHAN, Y. ZIGER-CFIR. 1977. Suppression of annuals by Artemisia herba-alba in the Negev desert of Israel. J. Ecol. 65: 413-426.

19. COWLES, H.C. 1911. The causes of vegetative cycles. Bot. Gaz. 51: 161-183.

20. BOOTH, W.E. 1941. Revegetation of abandoned fields in Kansas and Oklahoma. Am. J. Bot. 28: 415-422.

21. RICE, E.L., W.T. PENFOUND, L.M. ROHRBAUGH. 1960. Seed dispersal and mineral nutrition in succession in abandoned fields in central Oklahoma. Ecology 41: 224-228.

22. KAPUSTKA, L.A., E.L. RICE. 1976. Acetylene reduction (N_2-fixation) in soil and old field succession in central Oklahoma. Soil Biol. Biochem. 8: 497-503.

23. RICE, E.L. 1964. Inhibition of nitrogen-fixing and nitrifying bacteria by seed plants. I. Ecology 45: 824-837.

24. RICE, E.L., S.K. PANCHOLY. 1972. Inhibition of nitrification by climax ecosystems. Am. J. Bot. 59: 1033-1040.

25. RICE, E.L., S.K. PANCHOLY. 1973. Inhibition of nitrification by climax ecosystems. II. Additional evidence and possible role of tannins. Am. J. Bot. 60: 691-702.

26. RICE, E.L., S.K. PANCHOLY. 1974. Inhibition of nitrification by climax ecosystems. III. Inhibitors other than tannins. Am. J. Bot. 61: 1095-1103.

27. JACKSON, J.R., R.W. WILLEMSEN. 1976. Allelopathy in the first stages of secondary succession on the piedmont of New Jersey. Am. J. Bot. 63: 1015-1023.

28. WALTERS, D.T., A.R. GILMORE. 1976. Allelopathic
    effects of fescue on the growth of sweetgum. J.
    Chem. Ecol. 2: 469-479.
29. HORSLEY, S.B. 1977. Allelopathic inhibition of black
    cherry by fern, grass, goldenrod, and aster. Can.
    J. Forest Res. 7: 205-216.
30. TUBBS, C.H. 1973. Allelopathic relationship between
    yellow birch and sugar maple seedlings. For. Sci.
    19: 139-145.
31. JOBIDON, R., J.R. THIBAULT. 1981. Allelopathic
    effects of balsam poplar on green alder germination.
    Bull. Torrey Bot. Club 108: 413-418.
32. JOBIDON, R., J.R. THIBAULT. 1982. Allelopathic
    growth inhibition of nodulated and unnodulated
    Alnus crispa seedlings by Populus balsamifera.
    Am. J. Bot. 69: 1213-1223.
33. HANDLEY, W.R.C. 1963. Mycorrhizal associations and
    Calluna heathland afforestation. Bull. Forest
    Commn. London, No. 36.
34. ROBINSON, R.K. 1972. The production by roots of
    Calluna vulgaris of a factor inhibitory to growth
    of some mycorrhizal fungi. J. Ecol. 60: 219-224.
35. BROWN, R.T., P. MIKOLA. 1974. The influence of
    fruticose soil lichens upon the mycorrhizae and
    seedling growth of forest trees. Acta Forest Fenn.
    141: 1-22.
36. McCALLA, T.M., F.L. DULEY. 1948. Stubble mulch
    studies: Effect of sweetclover extract on corn
    germination. Science 108: 163.
37. McCALLA, T.M., F.L. DULEY. 1949. Stubble mulch
    studies: III. Influence of soil microorganisms and
    crop residues on the germination, growth and direc-
    tion of root growth of corn seedlings. Proc. Soil
    Sci. Soc. Am. 14: 196-199.
38. KATZNELSON, J. 1972. Studies in clover soil sickness.
    I. The phenomenon of soil sickness in berseem and
    Persian clover. Plant and Soil 36: 379-393.
39. TAMURA, S., C. CHANG, A. SUZUKI, S. KUMAI. 1967.
    Isolation and structure of a novel isoflavone
    derivative in red clover. Agric. Biol. Chem. 31:
    1108-1109.
40. TAMURA, S., C. CHANG, A. SUZUKI, S. KUMAI. 1969.
    Chemical studies on "clover sickness" Part I.
    Isolation and structural elucidation of two new
    isoflavonoids in red clover. Agric. Biol. Chem.
    33: 391-397.

41. CHANG, C.F., A. SUZUKI, S. KUMAI, S. TAMURA. 1969.
    Chemical studies on "clover sickness" Part II.
    Biological functions of isoflavonoids and their
    related compounds. Agric. Biol. Chem. 33: 398-408.
42. CHOU, C.H., H.J. LIN. 1976. Autointoxication mecha-
    nisms of Oryza sativa. I. Phytotoxic effects of
    decomposing rice residues in soil. J. Chem. Ecol.
    2: 353-367.
43. RICE, E.L., C.Y. LIN, C.Y. HUANG. 1981. Effects of
    decomposing rice straw on growth of and nitrogen
    fixation by Rhizobium. J. Chem. Ecol. 7: 333-344.
44. BURGOS-LEON, W., F. GANRY, R. NICOU, J.L. CHOPART, Y.
    DOMMERGUES. 1980. Etudes et travaux: un cas de
    fatigue des sols induite par la culture du sorgho.
    Agron. Trop. 35: 319-334.
45. OLIVER, L.R. 1979. Influence of soybean (Glycine max)
    planting date on velvetleaf (Abutilon theophrasti)
    competition. Weed Sci. 27: 183-188.
46. STANIFORTH, D.W. 1965. Competitive effects of three
    foxtail species on soybeans. Weeds 13: 191-193.
47. HAGOOD, E.S. JR., T.T. BAUMAN, J.L. WILLIAMS JR., M.M.
    SHREIBER. 1980. Growth analysis of soybeans
    (Glycine max) in competition with velvetleaf
    (Abutilon theophrasti). Weed Sci. 28: 729-734.
48. CHANDLER, J.M. 1977. Competition of spurred anoda,
    velvetleaf, prickly sida, and Venice mallow in
    cotton. Weed Sci. 25: 151-158.
49. ROBINSON, E.L. 1976. Effect of weed species and
    placement on seed cotton yields. Weed Sci. 24:
    353-355.
50. ELMORE, C.D. 1980. Inhibition of turnip (Brassica
    rapa) seed germination by velvetleaf (Abutilon
    theophrasti) seed. Weed Sci. 28: 658-660.
51. COLTON, C.E., F.A. EINHELLIG. 1980. Allelopathic
    mechanisms of velvetleaf (Abutilon theophrasti
    Medic., Malvaceae) on soybean. Am. J. Bot. 67:
    1407-1413.
52. BHOWMIK, P.C., J.D. DOLL. 1979. Evaluation of allelo-
    pathic effects of selected weed species on corn and
    soybeans. Proc. North Central Weed Control Conf.
    34: 43-45.
53. BHOWMIK, P.C., J.D. DOLL. 1982. Corn and soybean
    response to allelopathic effects of weed and crop
    residues. Agron. J. 74: 601-606.
54. HOLM, L. 1969. Weed problems in developing countries.
    Weed Sci. 17: 113-118.

55. FRIEDMAN, T., M. HOROWITZ. 1971. Biologically active
     substances in subterranean parts of purple nutsedge.
     Weed Sci. 19: 398-401.
56. HOROWITZ, M., T. FRIEDMAN. 1971. Biological activity
     of subterranean residues of Cynodon dactylon L.,
     Sorghum halepense L., and Cyperus rotundus L. Weed
     Res. 11: 88-93.
57. LUCENA, J.M., J. DOLL. 1976. Efectos inhibidores de
     crecimiento del coquito (Cyperus rotundus L.) sobre
     sorgo y soya. Revista Comalfi 3: 241-256.
58. PETERS, E.J. 1968. Toxicity of tall fescue to rape
     and birdsfoot trefoil seeds and seedlings. Crop
     Sci. 8: 650-653.
59. FAY, P.K., W.B. DUKE. 1977. An assessment of allelo-
     pathic potential in Avena germplasm. Weed Sci.
     25: 224-228.
60. LEATHER, G.R. 1983. Sunflowers (Helianthus annuus)
     are allelopathic to weeds. Weed Sci. 31: 37-42.
61. PUTNAM, A.R., J. DEFRANK. 1979. Use of cover crops
     to inhibit weeds. Proc. IX Int. Cong. Plant Protec-
     tion, pp. 580-582.
62. PUTNAM, A.R., J. DEFRANK. 1983. Use of phytotoxic
     plant residues for selective weed control. Crop
     Prot. 2: 173-181.
63. EVENARI, M. 1949. Germination inhibitors. Bot. Rev.
     15: 153-194.
64. PATRICK, Z.A. 1971. Phytotoxic substances associated
     with the decomposition in soil of plant residues.
     Soil Sci. 111: 13-18.
65. TANG, C.S., A.C. WAISS JR. 1978. Short-chain fatty
     acids as growth inhibitors in decomposing wheat
     straw. J. Chem. Ecol. 4: 225-232.
66. NORSTADT, F.A., T.M. McCALLA. 1963. Phytotoxic
     substance from a species of Penicillium. Science
     140: 410-411.
67. SPOEHR, H.A., J.H.C. SMITH, H.H. STRAIN, H.W. MILNER,
     G.J. HARDIN. 1949. Fatty Acid Antibacterials From
     Plants. Pub. 586. Carnegie Institution of Washing-
     ton, D.C.
68. ROBINSON, T. 1983. The Organic Constituents of Higher
     Plants. Fifth Edition. Cordus Press, North Amherst,
     Massachusetts, 353 pp.
69. KOBAYASHI, A., S. MORIMOTO, Y. SHIBATA, K. YAMASHITA, M.
     NUMATA. 1980. C_{10}-Polyacetylenes as allelopathic
     substances in dominants in early stages of secondary
     succession. J. Chem. Ecol. 6: 119-131.

70. CAMPBELL, G., J.D.H. LAMBERT, T. ARNASON, G.H.N. TOWERS. 1982. Allelopathic properties of α-terthienyl and phenylheptatriyne, naturally occurring compounds from species of Asteraceae. J. Chem. Ecol. 8: 961-972.

71. WHITTAKER, R.H., P.P. FEENY. 1971. Allelochemics: chemical interactions between species. Science 171: 757-770.

72. VARGA, M., E. KÖVES. 1959. Phenolic acids as growth and germination inhibitors in dry fruits. Nature 183: 401.

73. OWENS, L.D. 1969. Toxins in plant disease: structure and mode of action. Science 165: 18-25.

74. FENICAL, W. 1975. Halogenation in the Rhodophyta - A review. J. Phycol. 11: 245-259.

75. OWENS, L.D., J.F. THOMPSON, P.V. FENNESSEY. 1972. Dihydrorhizobitoxine, a new ether amino-acid from Rhizobium japonicum. J. Chem. Soc., Chem. Commu. 1972: 715.

76. RICE, E.L. 1983. Pest Control With Nature's Chemicals: Allelochemicals and Pheromones in Gardening and Agriculture. University of Oklahoma Press, Norman, Oklahoma.

77. AKAZAWA, T., P. MILJANICH, E.E. CONN. 1960. Studies on cyanogenic glycoside of Sorghum vulgare. Plant Physiol. 35: 535-538.

78. ABDUL-WAHAB, A.S., E.L. RICE. 1967. Plant inhibition by Johnson grass and its possible significance in old-field succession. Bull. Torrey Bot. Club 94: 486-497.

79. PATRICK, Z.A. 1955. The peach replant problem in Ontario. II. Toxic substances from microbial decomposition products of peach root residues. Can. J. Bot. 33: 461-486.

80. CHOU, C.H., G.R. WALLER. 1980. Possible allelopathic constituents of Coffea arabica. J. Chem. Ecol. 6: 643-654.

Chapter Five

PLANT ALLELOCHEMICALS: LINKAGES BETWEEN HERBIVORES AND
THEIR NATURAL ENEMIES

PEDRO BARBOSA

Department of Entomology
University of Maryland
College Park, Maryland 20742

JAMES A. SAUNDERS

U.S.D.A., A.R.S.
Plant Genetics and Germplasm Institute
Tobacco Laboratory
Beltsville, Maryland 20705

INTRODUCTION

Insect plant interactions have dominated the thoughts
and activities of plant biochemists, insect ecologists and
other entomologists for many years. A great deal of effort
has been devoted to the development of unifying principles
and general concepts that would be parsimonious with available empirical and experimental data and which would
accurately describe the patterns exhibited in the multitude
of insect-plant interactions.

Dethier[1] stated that "feeding preferences", and also
presumably host plant utilization patterns of insect herbivores, "resolve themselves into a dynamic equilibrium

between two changing systems, the plant and the insect".
In the 29 years since that publication most scientists have
conducted research as if they assumed that insect-plant
interrelations influence and are influenced only by the plant
or insect herbivore involved. Price et al.[2] and Bergman and
Tingey[3] have argued that theory on insect-plant interactions
cannot progress without careful consideration of the third
trophic level. Their conclusions were in large part based
on examples of allelochemical linkages among plants, herbivores and natural enemies. More importantly they proposed
that the third trophic level must be considered as part of a
plant's battery of defenses against herbivores. These argu-
ments depend on the assumption that plant defenses have a
significant favorable influence on natural enemies. Although
intuitively appealing, this assumption remains to be fully
and critically tested particularly as it applies to the
effects of allelochemicals within host tissues on parasitoids, pathogens and insect predators of insect herbivores.
However, various studies have provided some insight into
these issues as well as illustrating possible mechanisms
linking organisms in the three trophic levels.

Current theory on insect-plant interactions as well as
the application of the theory (e.g., in the breeding of
plant resistance to insect attack) has virtually ignored the
potential importance of natural enemies. However, the nature
and stability of plant-herbivore systems may be shaped or at
least influenced by natural enemies and indeed may be determined by the balance of offensive and defensive patterns
among plants, herbivores and natural enemies. Indeed, if
plant allelochemicals are found to have important roles in
the survival and effectiveness of natural enemies, this will
force modifications in current theories concerning both the
distribution of allelochemicals among plant species and
within plant individuals, and their role in plant defense.

Current theory on plant chemical defense has suggested
that although plants may be taxonomically and biologically
distinct there are, nevertheless, common developmental and
ecological characteristics which result in some degree of
convergence in their chemical defenses.[4-13] Other studies
have shown that biochemical constituents in plants vary
with time of day, the seasons, growth stage, the tissue,
climatic condition, soil condition, etc.[14-21] Consideration
of the selective forces which might be responsible for broad
patterns of defense, in which chemicals vary greatly in

distribution, has led to the hypothesis that such variation
in defensive chemistry reflects a strategy whereby tissues
that have the greatest value are protected. Value may be
defined in terms of photosynthetic capacity, nutrient
storage, the production of homeostatic messengers (e.g.,
hormones) or the involvement of tissues or organs in reproductive activities. The herbivore has costs similar to
those of plants and linked to its fitness, e.g. energy
allocated to physiological or behavioral defenses against
allelochemicals.[17,22-24] Differences in the quantity and/or
quality of allelochemicals in time or space, can create a
pattern of defense which is highly variable and against
which it is more difficult for the herbivore to adapt. The
question still remains, whether the theories on the relation
between the protection of "valuable" tissues and variable
defenses constitute general patterns of defense.

Nevertheless, ecologists have suggested that plants
should concentrate defensive compounds in valuable organs.
When one considers only the plant and the herbivore one
would expect herbivores to be found primarily on the less
defended plant parts. However, if herbivores can defend
themselves against natural enemies with these toxins then
specialist herbivores may concentrate their feeding on
organs high in allelochemicals. This would put selective
pressure on plants to keep toxins away from valuable organs.
Thus, predictions made that include natural enemies may be
opposite to those that exclude natural enemies from consideration, if indeed plant toxins do affect natural enemies.
Similarly, toxins in early successional plants may not be as
advantageous to those plants as currently assumed if specialist herbivores are able to utilize toxins to their advantage
against their own natural enemies. Digestibility reducing
compounds, for example in dominant trees, may only be useful
to the plant if natural enemies can take advantage of the
slower growth of herbivores on these plants and cause higher
mortality. Otherwise, herbivores left alone on such plants
would probably consume more foliage than they would on an
undefended plant (see[2]).

There are also some practical implications that should
be considered. Although we have known of chemical defenses
in plants since antiquity our use of this knowledge in
breeding plants resistant to insects is in its scientific
infancy. In the last fifteen years, knowledge of plant
chemistry has grown substantially. However, little if any,

emphasis has been placed on chemically mediated plant-
herbivore interactions which involve economic crop species,
pest species and their natural enemies.

The use of host plant resistance often relies on
increasing levels of allelochemicals responsible for antibiosis. Available data indicates that natural enemies like
parasitoids and predators are significantly more sensitive
to synthetic toxic compounds than their herbivore host (or
prey). In addition, our data (see below) and that of
others[3,25] suggests that although an increase in the concentration of a given allelochemical may increase the mortality
of an herbivore it may very well take a greater toll of its
parasitoids and thus be counterproductive to biological
control efforts.

PLANT ALLELOCHEMICALS AND INSECT PARASITOIDS

The critical role of plant allelochemicals in host-
habitat finding and host selection has been reviewed in
considerable detail.[26-30] Indeed, specific effects of
allelochemicals within the host on natural enemies may be
enhanced or counteracted by behavioral and biochemical
dynamics prior to parasitism. Nevertheless, this discussion
will focus solely on what occurs following parasitism as it
relates to the influence of allelochemicals.

The suitability of herbivores like the gypsy moth,
Lymantria dispar to their parasitoids is often affected by
the host plant consumed by the herbivore.[31-38] These host
plant associated changes alter the development, size and
survival of parasitoids of the gypsy moth.[39,40] There are
several other lines of evidence from various studies which
also suggest that plant allelochemicals and/or nutritional
differences within herbivore tissues may influence the
survival and development of herbivore parasitoids. However,
in most available studies no direct cause and effect relationships have been established between changes in diet-
mediated quality of herbivores and variation in the survival
and development of parasitoids, although numerous correlations are demonstrated and various hypotheses formulated.

A number of studies have noted that parasitoid development differs when the diet of the host herbivore changes.
It has been proposed that differential parasitoid development

may be due to an inappropriate balance or absence of key
nutritional compounds within the host which in turn reflect
differences in herbivore diets.[40-44] Greenblatt and
Barbosa[39] suggest that changes in parasitoids may be due to
host plant-mediated differences in the levels of herbivore
amino acids and carbohydrates. Still others speculate or
their work suggests that variation in plant allelochemicals
within host tissues and thus in host suitability may play a
major role in parasitoid development, survival and overall
fitness.[45-52] This type of effect and the impact of these
changes on the biology and ecology of parasitoids remain to
be fully explored since data from any one individual study
only provides one aspect of plant allelochemical-herbivore-
parasitoid interactions. Data on correlations between
herbivory of plants high in allelochemicals and the detrimental effects on parasitoid survival (or percent parasitism)
have often served as the basis for suggesting the importance
of allelochemicals in parasitoid fitness. Many of the
available studies suffer from, and future experimentation
must avoid, a series of problems which make data interpretation difficult (see discussion of allelochemicals and
predators).

Only a few studies have provided direct and compelling
evidence of either the influences of nutritional deficiencies
in herbivore diets on their parasitoids[53] or the effects of
allelochemicals on parasitoids. Morgan[54] and Gilmore[55,56]
provided data suggesting the possibility that nicotine might
influence the level of parasitism of the tobacco hornworm by
Apanteles congregatus. Thurston and Fox[57] were the first to
provide something other than circumstantial evidence. They
found a significantly lower percent larval parasitoid emergence from hosts treated with nicotine. Development time of
parasitoid larvae from nicotine-treated hornworms was longer,
but not significantly so. Campbell and Duffey[25] have shown
that tomatine (a major alkaloid of tomato) can cause prolonged
larval period, reduced pupal eclosion, smaller size and a
shortened adult longevity in Hyposoter exiguae, a parasitoid
of Heliothis zea (Fig. 1).

Campbell and Duffey's[25,58] studies were the first to
directly detect the allelochemical in the tissues of a
parasitoid and simultaneously demonstrate its deleterious
effects. Previous studies in which allelochemicals had been
detected failed to find any effects on the parasitoids or
provided what was, at best, circumstantial evidence of an

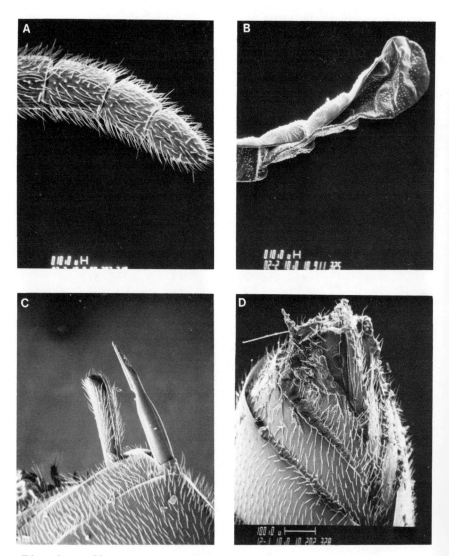

Fig. 1.   Effects of tomatine on morphology of Hyposoter
exiguae.   A. Antenna (control), B. Antenna (5% α-tomatine),
C. Female ovipositor (control), D. Female ovipositor (3%
tomatine).   (Photos courtesy of Dr. B. Campbell).

Table 1. Amount of nicotine in tissues and excretory
products of Cotesia congregata emerging from tobacco
hornworms (Manduca sexta) reared on a 0.1% nicotine diet.

	Adult	Cocoons	Meconium
No. of samples	37	37	35
Mean μg Nic./g.D.W.	8.1	126.5	82.0

Table 2. Amount of nicotine in tissues and excretory
products of Hyposoter annulipes emerging from armyworm
(Spodoptera frugiperda) reared on a 0.025% nicotine diet.

	Adult	Cocoons	Meconium
No. of samples	6	7	6
No. of indiv./sample	13	5	5
Mean μg Nic./g.D.W.	3.8	43.5	36.3

effect on parasitism.[47,50-52] Barbosa et al. (in manuscript) also provided direct evidence of the movement of
an allelochemical from food to herbivore to parasitoid
(Tables 1 and 2) in two different systems.

Studies on the tobacco hornworm, Manduca sexta, and
its parasitoid Cotesia congregata (= Apanteles congregatus)
and the fall armyworm, Spodoptera frugiperda, and its
parasitoid Hyposoter annulipes have demonstrated that, as
indicated in Thurston and Fox[57], there is a significantly
greater proportion of parasitoid larvae that fail to
emerge from hornworms reared on a nicotine diet (Table 3).
In addition, parasitism of nicotine treated hornworms
results in a higher proportion of parasitoid larvae that
fail to form cocoons (Table 3). No differences were noted
in developmental time or adult size (Table 4). Similar
results were noted for H. annulipes, the parasitoid of the

Table 3. Influence of nicotine on survival and development of Cotesia congregata, a parasitoid of Manduca sexta.

Treatment	Survival[1,2]					
	Total No. Larvae Produced	% Larvae[3] Failing to Emerge	% Larvae[3] Failing to Form Cocoons	% Pupal[3] Mort.	Total No. Larvae Emer.	Total No. Adults Emer.
Nicotine (0.1%)	93.0a	23.3a	19.9a	18.5a	67.7a	42.6a
Control	99.7a	8.1b	6.1b	15.5a	79.6a	54.6a

[1] All figures are least-squares estimates of means of 54 nicotine diet and 34 control diet reared hornworms.

[2] Values in columns with different letters are significant at $p < 0.05$ (SAS, Proc. GLM, Type III SS).

[3] All percentages were transformed to arc $\sin \sqrt{\%}$ prior to analysis. The values presented are the back-transformed mean percentages.

Table 4. Influence of nicotine on development of *Cotesia congregata*, a parasitoid of *Manduca sexta*.

Treatment	Development[1,2]			Size[1,2]
	Larval Devel. Time (Days)[3]	Pupal Devel. Time (Days)[4]	Total Devel. Time (Days)[5]	Dry Wt./Indiv. (mg)
Nicotine (0.1%)	11.91a	6.71a	17.71a	0.2240a
Control	11.73a	6.81a	17.52a	0.2443a

[1] All figures are least-squares estimates of means of 54 nicotine diet and 34 control diet reared hornworms.

[2] All values in columns with the same letters are not significantly different (SAS, Proc. GLM, Type III SS).

[3] Time period between oviposition and cocoon formation.

[4] Time period between cocoon formation and adult emergence.

[5] Sum total of time periods (2) and (3).

fall armyworm, although developmental parameters like
percent parasitism, larval development and adult size were
also detrimentally affected by nicotine (Table 5). Additional experimental support for these results was provided
by field studies in which high and low nicotine varieties
of tobacco were found to differentially affect the survival
and fitness of C. congregata (Thorpe and Barbosa, in
manuscript).

How do parasitoids cope with plant allelochemicals?
Barbosa et al. (in manuscript) traced nicotine to various
parasitoid tissues or materials secreted or excreted by
the parasitoid and demonstrated that both, C. congregata
and H. annulipes get rid of their toxin load by shunting
relatively high concentrations of nicotine into the cocoon
silk (produced by larvae) and into the meconium (which is
left behind) within the cocoon as adults emerge. The
compounds that actually occur in parasitoid tissues are
nicotine and cotinine. Although the tobacco hornworm
egests 98% of ingested nicotine,[59] the 1-2% found in their
blood is metabolized to cotinine (Fig. 2). The fall armyworm exhibits a similar mechanism in which large amounts
of ingested nicotine are excreted in the frass and the
remaining endogenous nicotine is metabolized to cotinine
(Fig. 3). Since it is unknown whether one or both alkaloids
are detrimental to parasitoid survival these data suggest
a caveat to those studying tri-trophic level interactions,
i.e., the plant allelochemical affecting an herbivore may
not be the toxin affecting its parasitoids.

The above studies not only provide evidence suggesting
that plant compounds may play key roles in host herbivore
suitability but more interestingly that allelochemically-
altered host quality can have a significant impact on
parasitoid survival and development, morphology, size,
percent parasitism, and perhaps on fecundity, percent
emergence, and efficiency of nitrogen utilization.

ALLELOCHEMICALS AND PATHOGENS

Plant allelochemicals may also mediate interactions
between herbivores and their pathogens. Allelochemicals
may directly inhibit pathogens which attack herbivores or
may act indirectly by altering herbivore susceptibility.
Indeed, other stresses have been shown to have a similar

Table 5. Influence of nicotine on survival and development of <u>Hyposoter annulipes</u>, a parasitoid of <u>Spodoptera frugiperda</u>.

Treatment	Survival			Development		Size	Sex Ratio
	% Larval Prod.	% Larval Failing to Form Cocoons	% Adult Emer.	Larval Devel. Time (Days)	Pupal Devel. Time (Days)	Dry Wt./Indiv. (mg)	% Females
Nicotine (0.025%)	46.4a	12.8a	74.2a	10.48a	6.34a	0.643a	48.4a
Control	75.8b	3.7b	81.3a	8.90b	6.29a	0.78b	41.3a

Values in columns with different letters are significantly different at $p < 0.05$.

PEDRO BARBOSA AND JAMES A. SAUNDERS

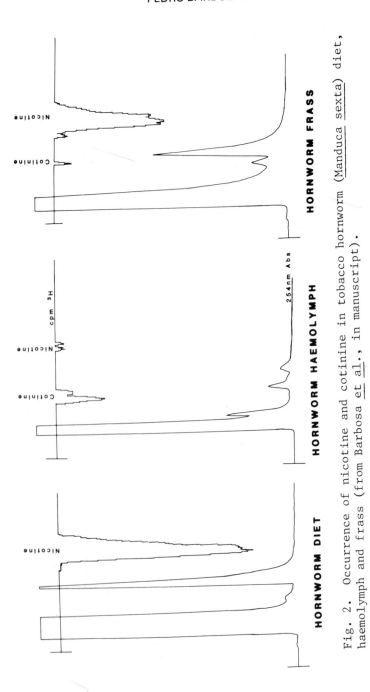

Fig. 2. Occurrence of nicotine and cotinine in tobacco hornworm (Manduca sexta) diet, haemolymph and frass (from Barbosa et al., in manuscript).

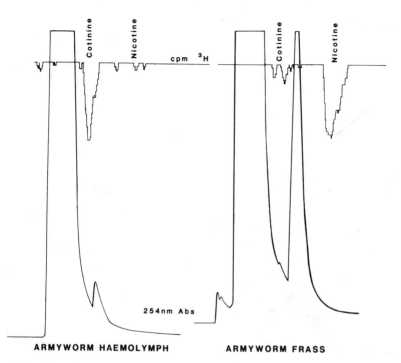

ARMYWORM HAEMOLYMPH   ARMYWORM FRASS

Fig. 3. Occurrence of nicotine and cotinine in fall
armyworm (Spodoptera frugiperda) haemolymph and frass.

effect.[60,61] While there is very little evidence of a
direct influence of plant allelochemicals on insect pathogens, there is some supportive evidence on the influence of
plant allelochemicals or plant extracts on plant pathogens
and encouraging preliminary evidence of the influence of
plant extracts on insect pathogens.

ALLELOCHEMICALS AND PLANT PATHOGENS

     Resistance rather than susceptibility of plants to
microorganisms appears to be the rule. This resistance is
in great part due to internal and external chemical defenses.
There are a variety of reports of unidentified fungitoxic
exudates on leaf surfaces which protect against plant
pathogens. Other types of biochemicals like tannins and

other phenolics act as non-specific microorganism inhibitors.[62] Leaf washes of elder and privet inhibit <u>Botrytis</u>
<u>cinnerea</u> spore germination and germ tube elongation.[63]
Similarly, toxic constituents of young cowpea leaf
diffusates inhibit germination of <u>Cercospora</u> (leaf spot)
conidia. These and other reports of fungitoxic effects of
leaf washings suggest that a variety of allelochemicals
play an important role in persistence of leaf surface
mycoflora and perhaps "specific" inhibition of foliar
pathogens. Roots may also exude toxic allelochemicals.
Rangaswami and Balasubramanian[64] showed that sorghum root
exudate (hydrocyanic acid) delayed germination of spores
of <u>Helminthosporium</u> <u>turcicum</u> and <u>Fusarium</u> <u>moniliforme</u>.
The majority of antifungal allelochemicals are glycosides,
e.g., cyanogenic and phenolic glycosides. Various other
types of antifungal allelochemicals also have been reported
among the saponins, unsaturated lactones and mustard oils.

Alkaloids may also play a defensive role against
microorganisms. For example, α-tomatine which occurs in
<u>Solanum</u> and <u>Lycopersicon</u> has fungistatic properties and
provides resistance in tomatoes to <u>Cortium</u> <u>rolfsii</u>[65] and
<u>Fusarium</u> <u>oxysporum</u>.[66] Cultivars of <u>L. pimpinellifolium</u> are
resistant to the causal agent of bacterial wilt <u>Pseudomonas</u>
<u>solanacearum</u>.[67] Other solanaceous alkaloids like solanine,
chaconine and solanidine inhibit the growth of the pathogen
<u>Alternaria solani</u>.[68] Mitscher <u>et al</u>.[69] also demonstrated
the antibacterial effects of various plant alkaloids.
<u>Mycospharella</u> blight-resistant varieties of <u>Cicer arietinum</u>
have more glandular hairs that secrete malic acid than
susceptible cultivars. High concentrations of malic acid
inhibit spore germination and growth of the pathogen.
Thus, these and other data strongly suggest that since
allelochemicals strongly influence plant pathogens, they
may have a similarly strong influence on other microorganisms which are pathogenic to herbivores.

ALLELOCHEMICALS AND INSECT PATHOGENS

Several studies have shown that plants can impart
protection to herbivores. The majority of these studies have
reported the antipathogenic properties of plant extracts
while a few demonstrate the effectiveness of specific
allelochemicals. Still other studies have demonstrated
that protection from pathogen-induced mortality can be

gained directly by consumption of certain plants although
not with other neutral plants. For example, consumption of
Ricinus medicinalis by the larvae of Agrotis ypsilon Rott.,
Laphygma exigua Hbn. and Spodoptera littoralis Boisd.
dramatically reduces mortality due to Bacillus thuringiensis
and B. entomocidus subtoxicus. Two other host plants,
Gossypium barbadense and Trifolium alexandrinum fail to
provide any protection from bacteria induced mortality.[70]
Similarly, S. littoralis, S. exigua, Pieris rapae and
Bombyx mori reared on foliage of plants possessing antibacterial activity suffered less mortality due to infection
by various B. thuringiensis strains.[71]

Smirnoff and Hutchison[72] evaluated 74 plant species
for their ability to inhibit growth of B. thuringiensis.
Results varied but fluids extracted from mascerated conifer
needles produced greater inhibition than those of other
woody or herbaceous plants. A similar study also found that
fluids from crushed leaves or leaf extracts of both deciduous and coniferous plants contained water soluble substances
that inhibited bacterial growth. Bacteria tested included
several strains of B. cereus and B. thuringiensis as well
as Pseudomonas aeruginosa and Serratia marcescens.[73]
Maksymiuk[74] reported antibacterial compounds from crushed
foliage of 22 species in the Pinaceae, Taxaceae, Fagaceae,
Salicaceae, Hamamelidaceae and Leguminosae which inhibited
the growth of various strains of B. thuringiensis. Analyses
of antibacterial fractions from pitch pine foliage provided
preliminary evidence that hydroxy carboxylic acids might
have been the active components.[74]

Direct evidence of the role of specific allelochemicals
in the inhibition of growth of insect pathogens is also
available (see Kubo chapter), although limited. Smirnoff[75]
was able to demonstrate that terpenoids of Abies balsamea
foliage like α- and β-pinene, limonene, phellandren,
fenchone and thujone exhibited a bacteriocidal effect on
entomopathogenic species like B. cereus, B. thuringiensis
and B. dendrolimus as well as other common bacterial like
Streptococcus faecalis, Escherichia coli, B. subtilis, etc.

Myrcene, β-caryophyllene and α-pinene (terpenoids) as
well as gossypol, catechin and cotton tannins (phenolics)
have shown high suppressant activity (in vitro) against
B. thuringiensis. Cotton plant allelochemicals like
gossypol, tannins, β-caryophyllene and gallic acid

suppressed growth of five bacteria species, isolated from
the gut of Anthonomous grandis, the boll weevil and against
B. thuringiensis. The concentrations of allelochemicals
tested were those typically found in cotton. Boll weevils
reared on a diet high in gossypol content exhibited a
decreased level of gut bacteria even though the allelochemical concentration did not significantly affect fecundity.[76]
Jones et al.[77] provide similar evidence of the inhibition of
growth of enteric microorganisms by 2-furaldehyde and
2-furoic acid. However, in these and similar studies
(see[78]) the microorganisms are airborne leaf surface
organisms or endosymbionts.

Merdan et al.[71] noted that among plants exhibiting
antibacterial effects the highest activity was associated
with Morus alba leaves and thus presumably some compounds in
the foliage. However, Iizuka et al.[79,80] indicated that the
antibacterial substances, i.e., protocatechuic acid and
p-hydroxybenzoic acid, were found in the feces and thus they
were components of digestive juices of the silkworm, B. mori.
Although these compounds were not found in mulberry leaves
Iizuka et al.[79] did suggest that their precursors might be
found in the foliage. The role of foliage components
remained unclear since the digestive juice of larvae starved
for less than 2 hours had a third major compound, caffeic
acid, in high enough concentration to depress the growth of
S. faecalis.[81] Finally, Koike et al.[81] and Iizuka[82] reported
that chlorogenic acid (a constituent of mulberry leaves) was
the compound from which caffeic acid was derived. At certain
concentrations and pH, chlorogenic acid also synergistically
enhanced the antibacterial activity of caffeic acid.

Bacteria are not the only microorganisms against
which foliar extracts and specific allelochemicals are
effective. Smelyanets[83] demonstrated that terpenoids like
α- and β-pinene, limonene and carene also inhibited the
development of fungal species extracted from the European
pine-shoot moth larvae, larger pine shoot beetles and from
their galleries. Antiviral agents also have been found
associated with host plant foliage of insects like the fall
webworm, Hyphantria cunea, and the gypsy moth, Lymantria
dispar.

Hayashiya[84] found that a gut protein of B. mori plus
an enzyme (a chlorophyllase-a) originating from its plant
host, mulberry, produced a red fluorescent protein which

acted as an antiviral agent in the midgut. The fall
webworm (H. cunea) was more resistant to nucleopolyhedrosis
virus when fed laboratory diet incorporating foliage of its
host compared to plain diet.[85] Rossiter[86] concluded that
pitch pine afforded gypsy moth larvae protection from
nucleopolyhedrosis virus because larvae reared from field
egg masses collected off pitch pine suffered half or less
the amount of pathogen mortality of those collected from
other hosts.

Very little research is available on the mechanisms by
which allelochemicals exert their influence on herbivores.
Allelochemicals might inhibit or stimulate microorganisms
or act indirectly by altering the susceptibility of herbivores. The antibiotic quality of plant allelochemicals
retained within the hemolymph of insect herbivores may
impart protection from pathogens. Indeed, bacteriocidal
quantities have been reported in the hemolymph of Oncopeltus
fasciatus, which feeds on seeds of asclepiad plants.[87]
Clearly these correlations must be made with care since the
blood of other asclepiad feeders does not exhibit similar
activity.[48] In addition, a variety of allelochemicals,
toxic to natural enemies of herbivores are produced de novo
even though they feed on toxin-containing plants. A
similar de novo origin may exist for antimicrobial compounds
in hemolymph. Finally, allelochemicals may cause the rupture
of gut epithelium and thus allow entry and development of
insect pathogens[88] (see references therein) (Fig. 4).

Thus, there is some preliminary evidence available to
support the hypothesis that allelochemicals may play an
important role in the pathogenicity of microorganisms
affecting insect herbivores as well as their parasitoids.
Nevertheless, many important issues need clarification and
critical questions must be answered.

ALLELOCHEMICALS AND INSECT PREDATORS

Attempts to understand the role of plant allelochemicals, within the tissues of prey, in the survival,
development and fitness of insect predators has been somewhat
problematic. First and foremost, the available evidence is
elemental. There are, for example, some reports of the
movement of plant allelochemicals through three trophic
levels as evidenced by the detection of allelochemicals in

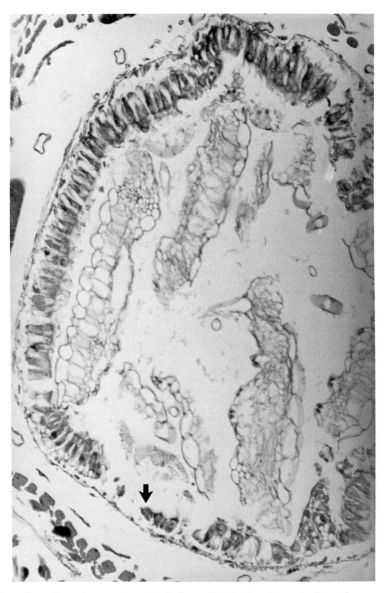

Fig. 4. Tannins extracted from <u>Liriodendron</u> <u>tulipifer</u> were
added to wild parsnip leaves, the normal host of the black
swallow tail caterpillar. This cross-section through
midgut of the second instar larva shows cellular detachment
of the basement membrane and accompanying necrosis of the
tissue. (Photo courtesy of Bruce Steinly.)

predators. However, very little preliminary data and no
conclusive evidence is available on the effects of the
transfer of plant allelochemicals to the predator trophic
level.

Hagen[89] reported that Adalia bipunctata readily feeds
and develops on elder-feeding Aphis sambuci. However,
larvae and adults of Coccinella septempunctata are adversely
affected by consumption of the same aphid species. Consumption of the aphid by the latter species of coccinellid
resulted in 100% mortality presumably due to the glycoside
in elder sambunigrin. Similarly, Pasteels[90] reported that
wing atrophication occurred in the coccinellid, Adonia
variegata when it fed on Aphis nerii feeding on the asclepiad, Cionura erecta. No abnormalities occurred when the
aphid prey had fed on another asclepiad, Cynachum acutum.
Progeny of wing atrophied females reared on A. nerii feeding
on C. erecta became wing atrophied adults while similar
progeny reared on a different aphid, Aphis craccivora from
a different plant, Robinia pseudoacacia, developed normally.
Although intuitively appealing and suggestive, the lack of
data for the role of plant allelochemicals in the development or mortality of the coccinellids fails to make this
proposal compelling.

Still other studies have only reported the presence
of plant allelochemicals in the extracts of predators
without reference to their biological or ecological import.
For example, cardenolides have been found in extracts of
Coccinella undecimpunctata L., a predator of the oleander-
feeding A. nerii. Similarly, cardenolides were present in
pupae of a lacewing Chrysopa sp. (presumably Carnea L.).[91]
However, other A. nerii feeding predators like C. septempunctata, an unidentified syrphid larva and a cecidomyiid larva
(presumed to be Aphidoletes aphidimyza) were found to be
negative in tests for cardenolides. This highlights one of
the problems in the interpretation of even the limited data
available. Lack of evidence of plant allelochemicals in
the tissues of a predator may mean that none were taken up
from its prey. Alternatively, the inability to detect an
allelochemical may be due to predator efficiency in egesting
the allelochemical (as in the tobacco hornworm, Manduca
sexta[92,93]), the lack of the allelochemical in the strain
of plant consumed by the prey (e.g., certain cardenolides
may be present or totally absent in different strains of
Asclepias curassavica) or the dramatic seasonal disappearance

of specific allelochemicals in the host plants of the prey
(e.g., Lotus corniculatus may be positive for HCN in spring
and negative in autumn).[94] On the other hand, the ability
to detect an allelochemical in a predator extract may reflect
the presence of the allelochemical in the gut contents (prey
bolus) of the predator rather than in predator tissues.
Obviously, the latter does not preclude the allelochemical
having a biologically significant effect.

When positive evidence of an allelochemical represents
the detection of a general chemical group, the specific
chemical present in the predator may actually be different
from that in its prey and may be produced de novo. For
example, puparia extracts of the parasitoid Zenillia adamsoni
were shown to have digitalis-like activity as were extracts
of its host but the specific compound in parasitoid extracts
was probably not the same as the one in its host. Thus, the
parasitoid compound may have been a different plant allelochemical or a product of de novo synthesis. Even when a
specific compound is identical to that found in a prey or
its host plant, the compound may still be produced de novo.
Consider species of the Lepidoptera genus Zygaena which
release HCN from body tissues whether reared on their cyanogenic host plants or not.[47]

Finally, the role of plant allelochemicals, in the
tissues of prey, may influence predator fitness through
their effects on predator behavior, i.e., predator avoidance
of certain allelochemical-containing prey. Jones[95] noted
that Lygaeus kalmii was extremely distasteful to ants.
Duffey[96] noted that L. kalmii which fed on North American
asclepiads contained cardiac glycosides. However, it is
difficult to draw any conclusions from these studies. Many
insects, including many Hemiptera, secrete repellent substances and thus it is certainly possible that, apart from
cardenolides, other toxic substances might be found in the
repugnatorial glands of insects like Lygaeid bugs.[96] Not
all insects have such glands and indeed there are a limited
number of reports of insect predator-protected herbivores,
which presumably sequester plant allelochemicals.[91,97]
These phenomena require more detailed evaluation of
predator-prey behavioral interactions induced by plant
allelochemicals.[98]

CONCLUSION

Plant allelochemicals can influence insect herbivores by affecting their physiology, behavior, development and survival. That is, allelochemicals can induce mortality or may act as behavioral modifiers (by altering orientation, mating, feeding, etc.) or physiological modifiers (by altering metabolic efficiency, growth, etc.). This review has provided some of the evidence indicating the potential roles of allelochemicals in analogous behavioral, physiological and ecological interactions between natural enemies and their herbivore hosts. Although more data is needed, the results of the studies on natural enemies discussed here as well as those discussed in other studies[2,26,27,29,99,100] on the role of allelochemicals on short and long range orientation illustrate how potentially pervasive may be the influence of allelochemicals.

If indeed plant allelochemicals have significant detrimental effects on natural enemies living in or on herbivore host which feed on "toxin" containing plants (TCP) (i.e., plants with allelochemicals that are potentially toxic) one might ask, "Do the benefits and drawbacks of feeding on TCP, for the herbivore, result in a zero-sum gain?". For the generalist herbivore, which presumably feeds on TCP at some cost, the benefits accrued if the plant toxins act as a selective force moving natural enemies towards the utilization of hosts feeding on non-toxic plants, may be counterbalanced by the effects on its own fitness. However, for the specialist, which presumably at very worst pays a small price for the use of TCP and may even require allelochemicals for growth and survival[101,102] the effects on parasitoids (or other natural enemies) may further enhance the overall benefits of the utilization of TCP.

For the generalist herbivore (or perhaps some specialist herbivores) the effects of TCP on their growth (i.e., the prolongation of the developmental period) may play a crucial role in the relative impact of allelochemicals on the biology and evolution of its natural enemies. For example, one might speculate that the prolonged developmental period of the larval herbivore feeding on TCP might broaden its window of vulnerability to parasitoids.[103] Such an enhanced opportunity for parasitism of hosts and presumed increase in parasitoid populations might counterbalance detrimental effects such as those noted in this chapter. Alternatively,

one might propose that the extended developmental period
of herbivore hosts merely increases the total exposure of
developing larval parasitoids to a detrimental toxin. These
and other alternatives may indeed not be mutually exclusive
since the nature of inter- and intratrophic level interactions such as these may depend on various aspects of the
life history of the natural enemies and/or the physiology
of the herbivore and toxicological properties of the plant
allelochemical.

Our understanding of tri-trophic level interactions of
the types discussed here is just in the formative stages.[104]
The questions that remain are numerous but intriguing and
the challenges exciting.

ACKNOWLEDGMENTS

Work described herein was supported by USDA competitive
Grant No. 59-2241-1-1-749-0. The authors would like to
express appreciation for the assistance, council and efforts
of Gail Barbosa, John Kemper, Joseph Olechno, Robert Trumbule
and Peter Martinat. The ideas, criticisms and suggestions
of Dr. Paul Gross are highly appreciated. Scientific Article
No. A-4041 of the Maryland Agricultural Experiment Station,
Department of Entomology.

REFERENCES

1. DETHIER, V.G. 1954. Evolution of feeding preferences
        in phytophagous insects. Evolution 8: 33-54.
2. PRICE, P.W., C.E. BOUTON, P. GROSS, B.A. McPHERSON,
        J.N. THOMPSON, A.E. WEIS. 1980. Interactions among
        three trophic levels. Influence of plants on inter-
        actions between insect herbivores and natural
        enemies. Annu. Rev. Ecol. Syst. 11: 41-65.
3. BERGMAN, J.M., W.M. TINGEY. 1979. Aspects of interac-
        tion between plant genotypes and biological control.
        Bull. Entomol. Soc. Amer. 25: 275-279.
4. FEENY, P. 1970. Seasonal changes in oak leaf tannins
        and nutrients as a cause of spring feeding by winter
        moth caterpillars. Ecology 57: 565-581.
5. FEENY, P. 1976. Plant apparency and chemical defense.
        In Biochemical Interactions Between Plants and
        Insects. Recent Advances in Phytochemistry.

(J.W. Wallace, R.L. Mansell, eds.), Vol. 10, Plenum Press, New York, pp. 1-14.

6. RHOADES, D.F., R.G. CATES. 1976. Toward a general theory of plant anti-herbivore chemistry. In J.W. Wallace, R.L. Mansell, eds., op. cit. Reference 5, pp. 168-213.

7. FUTUYMA, D.J. 1976. Food plant specialization and environmental unpredictability in Lepidoptera. Amer. Nat. 110: 285-292.

8. BERNAYS, E.A. 1978. Tannins: An alternative viewpoint. Entomol. Exp. Appl. 24: 44-53.

9. FUTUYMA, D.J., F. GOULD. 1979. Associations of plants and insects in a deciduous forest. Ecol. Monogr. 49: 33-50.

10. BRATTSTEN, L. 1979. Biochemical defense mechanisms in herbivores against plant allelochemicals. In Herbivores. (G.A. Rosenthal, D.H. Janzen, eds.), Academic Press, New York, pp. 200-270.

11. CATES, R.G. 1980. Feeding patterns of monophagous, oligophagous, and polyphagous insect herbivores: The effect of resource abundance and plant chemistry. Oecologia 46: 22-31.

12. CATES, R.G. 1981. Host plant predictability and the feeding patterns of monophagous, oligophagous and polyphagous insect herbivores. Oecologia 48: 319-326.

13. AHMAD, S. 1982. Roles of mixed-function oxidases in insect herbivory. In Proc. 5th Intl. Symp. Insect-Plant Relationships. C.A.P.D., Wageningen, The Netherlands, pp. 41-47.

14. JONES, D.A. 1962. Selective eating of the acyanogenic forms of the plant Lotus corniculatus L. by various animals. Nature 193: 1109-1111.

15. ZAVARIN, E., L. LAWRENCE, M. THOMAS. 1971. Compositional variations of leaf monoterpenes in Cupressus macrocarpa, C. pygmaea, C. goveniana, C. abramisiana and C. sargentii. Phytochemistry 10: 379-393.

16. BROWER, L.P., M. EDMUNDS, C.M. MOFFITT. 1975. Cardenolide content and palatability of a population of Danaus chrysippus butterflies from W. Africa. J. Entomol. 49: 183-196.

17. RHOADES, D.F. 1977. The antiherbivore chemistry of larrea. In Creosote Bush: Biology and Chemistry of Larrea in New World Deserts. (T.J. Mabry, J.H. Hunziker, D.R. DiFeo, Jr., eds.), Dowden, Hutchinson and Ross, Inc., Stroudsburg, Pennsylvania.

18. SEIGLER, D.S., E.E. CONN, J.E. DUNN, D.H. JANZEN.
    1979. Cyanogenesis in Acacia farnesiana.
    Phytochemistry 18: 1389-1390.
19. LAGENHEIM, J.H., D.E. LINCOLN, W.H. STUBBLEBINE,
    A.C. GABRIELLI. 1982. Evolutionary implications
    of leaf resin pocket patterns in the tropical
    tree Hymenaea (Caesalpinioideae: Leguminosae).
    Amer. J. Bot. 69: 595-607.
20. SCHULTZ, J.C., P.J. NOTHNAGLE, I.T. BALWIN. 1982.
    Seasonal and individual variation in leaf quality
    of two northern hardwood tree species. Amer. J.
    Bot. 69: 753-759.
21. ZUCKER, W.V. 1982. How aphids choose leaves: the
    roles of phenolics in host selection by a galling
    aphid. Ecology 63: 972-981.
22. MOONEY, H.A., C. CHU. 1974. Seasonal carbon
    allocation in Heteromeles arbutifolia, a
    California evergreen shrub. Oecologia 49: 50-55.
23. CHEW, F.S., J.E. RODMAN. 1979. Plant resources
    for chemical defense. In G.A. Rosenthal, D.H.
    Janzen, eds., op. cit. Reference 10, pp. 271-
    307.
24. McKEY, D. 1979. The distribution of secondary
    compounds within plants. In G.A. Rosenthal,
    D.H. Janzen, eds., op. cit. Reference 10, pp.
    56-133.
25. CAMPBELL, B.C., S.S. DUFFEY. 1979. Tomatine and
    parasitic wasps: potential incompatibility of
    plant antibiosis with biological control. Science
    205: 700-702.
26. VINSON, S.B. 1975. Biochemical coevolution between
    parasitoids and their hosts. In Evolutionary
    Strategies of Parasitic Insects and Mites.
    (P.W. Price, ed.), Plenum Press, New York, pp.
    14-48.
27. VINSON, S.B. 1976. Host selection by insect parasi-
    toids. Annu. Rev. Entomol. 21: 109-133.
28. VINSON, S.B. 1977. Behavioral chemicals in the
    augmentation of natural enemies. In Biological
    Control by Augmentation of Natural Enemies.
    (R.L. Ridgway, S.B. Vinson, eds.), Plenum Press,
    New York, pp. 237-279.
29. VINSON, S.B. 1981. Habitat location. In Semio-
    chemicals. Their Role in Pest Control. (D.A.
    Nordlund, R.L. Jones, W.J. Lewis, eds.), John
    Wiley and Sons, Inc., New York, pp. 51-78.

30. PRICE, P.W. 1981. Relevance of ecological concepts
    to practical biological control. In BARC Symposium
    V. Biological Control in Crop Protection, Allanheld,
    Osmun, Publications, Totowa, New Jersey, pp. 3-19.
31. CAPINERA, J.L., P. BARBOSA. 1976. Dispersal of first-
    instar gypsy moth larvae in relation to population
    quality. Oecologia 26: 55-60.
32. CAPINERA, J.L., P. BARBOSA. 1977. Influence of
    natural diets and density on gypsy moth egg mass
    characteristics. Can. Entomol. 101: 1313-1318.
33. BARBOSA, P., J. GREENBLATT, W. WITHERS, W. CRANSHAW,
    E. HARRINGTON. 1979. Host plant preferences and
    their induction in larval of the gypsy moths,
    Lymantria dispar. Entomol. Exp. Appl. 26: 180-188.
34. BARBOSA, P., J.L. CAPINERA. 1977. The influence of
    food type in the developmental structure of labora-
    tory populations of the gypsy moth, Porthetria
    dispar L. Can. J. Zool. 55: 1424-1429.
35. BARBOSA, P., J.L. CAPINERA. 1978. Population quality,
    dispersal and numerical change in the gypsy moth,
    Lymantria dispar. Oecologia 36: 203-209.
36. BARBOSA, P. 1978. Host plant exploitation by the
    gypsy moth, Lymantria dispar L. Entomol. Exp. Appl.
    24: 228-237.
37. BARBOSA, P., J. GREENBLATT. 1979. Suitability,
    digestibility and assimilation of various host plants
    of the gypsy moth, Lymantria dispar L. Oecologia
    43: 111-119.
38. BARBOSA, P., W. CRANSHAW, J.A. GREENBLATT. 1981.
    Influence of food quality on polymorphic dispersal
    behaviors in the gypsy moth, Lymantria dispar.
    Can. J. Zool. 59: 293-297.
39. GREENBLATT, J.A., P. BARBOSA. 1981. Effects of host's
    diet on two pupal parasitoids of the gypsy moth:
    Brachymeria intermedia (Nees) and Coccygomimus
    turionellae (L.). J. Appl. Ecol. 18: 1-10.
40. GREENBLATT, J.A., P. BARBOSA, M.E. MONTGOMERY. 1982.
    Host's diet effects on nitrogen utilization effi-
    ciency for two parasitoid species, Brachymeria
    intermedia and Coccygomimus turionellae. Physiol.
    Entomol. 7: 263-267.
41. FLANDERS, S.E. 1942. Abortive development in parasitic
    Hymenoptera, induced by the food plant of the insect
    host. J. Econ. Entomol. 35: 834-835.

42. LANGE, R., J.F. BRONSKILL. 1964. Reactions of Musca domestica to parasitism by Aphaereta pallipes with special reference to host diet and parasitoid toxin. Z.f. Parasit. 25: 193-210.

43. PIMENTEL, D. 1966. Wasp parasite (Nasionia vitripennis) survival on its house fly host (Musca domestica) reared on various foods. Annu. Entomol. Soc. Amer. 59: 1031-1038.

44. CHENG, L. 1970. Timing of attack by Lypha dubia Fall. (Diptera: Tachinidae) on the winter moth, Operophthera brumata (Lepidoptera: Geometridae) as a factor affecting parasite success. J. Anim. Ecol. 39: 313-320.

45. NARAYANAN, E.S., B.R. SUBBA RAO. 1955. Studies in insect parasitism I-III. The effects of different hosts on the physiology, on the development and behaviour and on the sex ratio of Microbracon gelechiae Ashmead (Hymenoptera: Braconidae). Beit. Entomol. 5: 36-60.

46. SMITH, J.M. 1957. Effects of the food of California red scale, Aonidiella aurantii (Mask.) on reproduction of its hymenopterous parasites. Can. Entomol. 89: 219-230.

47. JONES, D.A., J. PARSONS, M. ROTHSCHILD. 1962. Release of hydrocyanic acid from crushed tissues of all stages in the life cycle of species of the Zygaeninae (Lepidoptera). Nature 193: 52-53.

48. REICHSTEIN, T., J. VON EUW, J.A. PARSONS, M. ROTHSCHILD. 1968. Heart poisons in the monarch butterfly. Science 161: 861-866.

49. ALTAHTAWY, M.M., S.M. HAMMAD, E.M. HEGAZI. 1976. Studies on the dependence of Microplitis rufiventris Kok. (Hym., Braconidae) parasitizing Spodoptera littoralis (Boisd.) on own food as well as on food of its host. Z. Ang. Entomol. 83: 3-13.

50. ROTHSCHILD, M., G. VALADON, R. MUMMERY. 1977. Carotenoids of the pupae of the large white butterfly (Pieris brassicae) and the small white butterfly (Pieris rapae). J. Zool. 181: 323-339.

51. SMITH, D.A.S. 1978. Cardiac glycosides in Danaus chrysippus (L.) provide some protection against an insect parasitoid. Experientia 34: 844-846.

52. BENN, M., J. DeGRAVE, C. GNANASUNDERAM, R. HUTCHINS. 1979. Host-plant pyrrolizidine alkaloids in Nyctemera annulata Boisd: their persistence through the life cycle and transfer to a parasite. Experientia 35: 731-732.

53. ZOHDY, N. 1976. On the effect of the food of Myzus
    persicae Sulz. on the hymenopterous parasite,
    Aphelinus asychis Walker. Oecologia 26: 185-191.
54. MORGAN, A.C. 1910. Observations recorded at the
    236th regular meeting of the Entomological Society
    of Washington. Proc. Entomol. Soc. Wash. 12: 72.
55. GILMORE, J.U. 1938. Observations on the hornworms
    attacking tobacco in Tennessee and Kentucky. J.
    Econ. Entomol. 31: 706-712.
56. GILMORE, J.U. 1938. Notes on Apanteles congregatus
    (Say) as a parasite in tobacco hornworms. J. Econ.
    Entomol. 31: 712-715.
57. THURSTON, R., P.M. FOX. 1972. Inhibition by nicotine
    of emergence of Apanteles congregatus from its host,
    the tobacco hornworm. Annu. Entomol. Soc. Amer.
    65: 547-550.
58. CAMPBELL, B.C., S.S. DUFFEY. 1981. Alleviation of
    α-tomatine-induced toxicity to the parasitoid,
    Hyposoter exiguae by phytosterols in the diet of
    the host Heliothis zea. J. Chem. Ecol. 7: 927-
    946.
59. SELF, L.S., F.E. GUTHRIE, E. HODGSON. 1964. Adapta-
    tions of tobacco hornworms to the ingestion of
    nicotine. J. Insect Physiol. 10: 907-914.
60. STEINHAUS, E.A., J.P. DINEEN. 1960. Observations on
    the role of stress in a granulosis of the variegated
    cutworm. J. Insect Pathol. 2: 55-65.
61. SMITH, K.M. 1976. Virus-Insect Relationships.
    Longman Group Limited Publishers, London, 291 pp.
62. HARBORNE, J.B. 1982. Introduction to Ecological
    Chemistry. 2nd Edition, Academic Press, New York,
    278 pp.
63. TOPPS, J.H., R.L. WAIN. 1957. Fungistatic properties
    of leaf exudates. Nature 179: 652-653.
64. RANGASWAMI, G., A. BALASUBRAMANIAN. 1963. Release of
    hydrocyanic acid by sorghum roots and its influence
    on the rhizosphere microflora and plant pathogen
    fungi. Ind. J. Exp. Biol. 1: 215-217.
65. SCHLOSSER, E. 1976. Role of saponins in antifungal
    resistance. VII. Significance of tomatine in
    species-specific resistance of tomato fruits against
    fruit rotting fungi. Meded. Fac. Landbouwwet.
    Ryksuniv. Gent. 41: 499-503.
66. IRVING, G.W., T.D. FONTAINE, S.P. DOOLITTLE. 1945.
    Lycopersicin, a fungistatic agent from the tomato
    plant. Science 102: 9-11.

67. MOHANAKUMARAN, N., J.C. GILBERT, I.W. BUDDENHAGEN. 1969. Relationship between tomatine and bacterial wilt resistance in tomato. Phytopathology 59: 14.

68. SINDEN, S.L., J.M. SCHALK, A.K. STONER. 1978. Effects of daylength and maturity of tomato plants on tomatine content and resistance to the Colorado potato beetle. J. Amer. Soc. Hort. Sci. 103: 596-600.

69. MITSCHER, L.A., R.P. LEU, M.S. BATHALA, W.N. WU, J.L. BEAL, R. WHITE. 1972. Antimicrobial agents from higher plants. I. Introduction, rationale and methodology. Lloydia. 35: 157-166.

70. AFIFY, VON A.M., A.I. MERDAN. 1969. Reaktionsunterschiede von drei Noctuiden-Arten bei bestimmten Bacilluspräparaten in Abhängigkeit von der Nahrung und Art der Behandlung. Anz. Schädlingskde. u. Pflanzenschutz 42: 102-104.

71. MERDAN, A., H. ABDEL-RAHMAN, A. SOLIMAN. 1975. On the influence of host-plants on insect resistance to bacterial diseases. Z. Ang. Entomol. 78: 280-285.

72. SMIRNOFF, W.A., P.M. HUTCHINSON. 1965. Bacteriostatic and bacteriocidal effects of extracts of foliage from various plant species on Bacillus thuringiensis var. thuringiensis Berliner. J. Invert. Pathol. 7: 273-280.

73. KUSHNER, D.S., G.T. HARVEY. 1962. Antibacterial substances in leaves: their possible role in insect resistance to diseases. J. Insect Pathol. 4: 155-184.

74. MAKSYMIUK, B. 1970. Occurrence in nature of antibacterial substances in plants affecting Bacillus thuringiensis and other enterobacteria. J. Invert. Pathol. 15: 365-371.

75. SMIRNOFF, W.A. 1972. Effects of volatile substances released by foliage of Abies balsamea. J. Invert. Pathol. 19: 32-35.

76. HEDIN, P.A., O.H. LINDIG, P.P. SIKOROWSKI, M. WYATT. 1978. Suppressants of gut bacteria in the boll weevil from the cotton plant. J. Econ. Entomol. 71: 394-396.

77. JONES, C.G., J.R. ALDRICH, M.S. BLUM. 1981. Baldcypress allelochemicals and the inhibition of silkworm enteric microorganisms. Some ecological considerations. J. Chem. Ecol. 7: 103-114.

78. JONES, C.G. 1984. Microorganisms as mediators of plant resource exploitation by insect herbivores. In A New Ecology: Novel Approaches to Interactive Systems. (P.W. Price, C.N. Slobodchikoff, W.S. Gaud), John Wiley and Sons, Inc., New York, pp. 53-99.
79. IIZUKA, T., S. KOIKE, J. MIZUTANI. 1974. Antibacterial substances in feces of silkworm larvae reared on mulberry leaves. Agric. Biol. Chem. 38: 1549-1550.
80. IIZUKA, T., S. KOIKE, J. MIZUTANI. 1975. Antibacterial activity of protocatechuic acid and p-hydroxybenzoic acid isolated from feces. J. Seric. Sci. 44: 125-130.
81. KOIKE, S., T. IIZUKA, J. MIZUTANI. 1979. Determination of caffeic acid in the digestive juice of silkworm larvae and its antibacterial activity against the pathogenic Streptococcus faecalis AD-4. Agric. Biol. Chem. 43: 1727-1731.
82. IIZUKA, T. 1983. Studies on the bacterial flora in the midgut and on the antibacterial activity in the digestive juice of larvae of the silkworm, Bombyx mori L. Sericologia 23: 227-244.
83. SMELYANETS, V.P. 1977. Mechanisms of plant resistance in scotch pine (Pinus sylvestris). 4. Influence of food quality on physiological state of pine pests (trophic preferendum). Z. Ang. Entomol. 84: 232-241.
84. HAYASHIYA, K. 1978. Red fluorescent protein in the digestive juice of the silkworm larvae fed on host plant mulberry leaves. Entomol. Exp. Appl. 24: 228-236.
85. KUNIMI, Y., H. ARUGA. 1974. Susceptibility to infection with nuclear and cytoplasmic polyhedrosis virus of the fall webworm, Hyphantria cunea Drury reared in several artificial diets. Jap. J. App. Entomol. Zool. 18: 1-4.
86. ROSSITER, M.A. 1981. Factors contributing to host range extension in the gypsy moth, Lymantria dispar. Ph.D. Dissertation. State University of New York.
87. FRINGS, H., E. GOLDBERG, J.C. ARENTZEN. 1948. Antibacterial action of the blood of the large milkweed bug. Science 108: 689-690.
88. BERENBAUM, M.R. 1983. Effects of tannins on growth and digestion in two species of papilionids. Entomol. Exp. Appl. 34: 245-250.

89. HAGEN, K.S. 1962. Biology and ecology of predaceous
    Coccinellidae. Annu. Rev. Entomol. 7: 289-326.
90. PASTEELS, J.M. 1978. Apterous and brachypterous
    coccinellids at the end of the food chain, Cionura
    erecta (Ascelepiadaceae) Aphis nerii. Entomol.
    Exp. Appl. 24: 579-584.
91. ROTHSCHILD, M., J. VON EUW, T. REICHSTEIN. 1973.
    Cardiac glycosides in a scale insect (Aspidiotus),
    a ladybird (Coccinella) and a lacewing (Chrysopa).
    J. Entomol. 48: 89-90.
92. SELF, L.S., F.E. GUTHRIE, E. HODGSON. 1964. Adaptation
    of tobacco hornworms to the ingestion of nicotine.
    J. Insect Physiol. 10: 907-914.
93. SELF, L.S., F.E. GUTHRIE, E. HODGSON. 1964. Metabolism
    of nicotine by tobacco feeding insects. Nature 204:
    300-301.
94. ROTHSCHILD, M., J. VON EUW, T. REICHSTEIN. 1970.
    Cardiac glycosides in the oleander aphid Aphis nerii.
    J. Insect Physiol. 16: 1141-1145.
95. JONES, F.M. 1932. Insect coloration and relative
    acceptability of insects to birds. Trans. R.
    Entomol. Soc. 80: 345-385.
96. DUFFEY, S.S. 1970. Cardiac glycosides and distaste-
    fulness: some observations on the palatability
    spectrum of butterflies. Science 169: 78-79.
97. URGUHART, F.A. 1960. The Monarch Butterfly. Univ.
    of Toronto Press, Toronto, Canada.
98. BERENBAUM, M.R. 1984. Mantids and milkweed bugs:
    efficacy of aposematic coloration against inverte-
    brate predators. Amer. Midl. Nat. 111: 64-68.
99. ARTHUR, A.P. 1981. Host acceptance by parasitoids.
    In D.A. Nordlund, R.L. Jones, W.J. Lewis, eds.,
    op. cit. Reference 29, pp. 97-120.
100. ELZEN, G.W., H.J. WILLIAMS, S.B. VINSON. 1983.
    Response by the parasitoid Campoletis sonorensis
    (Hymenoptera: Ichneumonidae) to chemicals (Synonomes)
    in Plants: implications for host habitat location.
    Environ. Entomol. 12: 1873-1877.
101. ROSENTHAL, G.A., D.H. JANZEN, D.L. DAHLMAN. 1977.
    Degradation and detoxification of canavanine by
    a specialized seed predator. Science 196: 658-660.
102. BERNAYS, E.A., S. WOODHEAD. 1982. Plant phenols
    utilized as nutrients by a phytophagous insect.
    Science 216: 201-203.

103. SCHULTZ, J.C. 1983. Impact of variable plant defen-
     sive chemistry on susceptibility of insects to
     natural enemies. In Plant Resistance to Insects.
     (P.A. Hedin, ed.), Amer. Chem. Soc., Washington,
     D.C., pp. 37-54.
104. BARBOSA, P., J.A. SAUNDERS, M. WALDVOGEL. 1982.
     Plant mediated variation in herbivore suitability
     and parasitoid fitness. In op. cit. Reference 13,
     pp. 63-71.

Chapter Six

BREMENTOWN REVISITED: INTERACTIONS AMONG ALLELOCHEMICALS
IN PLANTS

MAY BERENBAUM

Department of Entomology
University of Illinois
Urbana, Illinois 61801

"When we are all performing together it will have a very
good effect." The ass, in "The Brementown Musicians"
(J. and W. Grimm, 1814)

INTRODUCTION

While certain individual plant chemicals are closely
identified with particular plant families—sinigrin and the
Cruciferae, for example[1]—plant species in reality produce
a variety of biosynthetically distinct secondary compounds.
The tremendous chemical diversity that characterizes the
angiosperms in general holds for an individual plant as well.
The Umbelliferae are among the more diverse; even the rather
undistinguished parsnip (Pastinaca sativa), a biennial that
is a minor vegetable crop under cultivation and a noxious
weed when naturalized, contains no fewer than seven classes

139

Table 1. Secondary compounds identified from _Pastinaca sativa_.

Coumarins[88]	Terpenes[62,104]	Flavonoids[105]
Coumarin	α-Thujene	Rutin
Osthol	Camphene	Hyperin
Umbelliferone	β-Pinene	Isorhamnetin diglycoside
	Sabinene	Isoquercitin
Furanocoumarins[88,106,107]	Myrcene	Isorhamnetin-3-glucoside
Bergapten	α-Phellandrene	Isorhamnetin-3-glucoside-7-rhamnoside
Imperatorin	Limonene	
Apterin	β-Phellandrene	Polyacetylenes[82,105]
Isobergapten	β-Ocimene	Falcarinol
Isopimpinellin	γ-Terpinene	Falcarindiol
Pimpinellin	p-Cymene	$H_2C=CHCO-[C{\equiv}C]_2CH_2CH=CH-[CH_2]_7CHO$
Sphondin	Terpinolene	
Xanthotoxin	cis-Allo-ocimene	
Xanthotoxol	β-Bisabolene	
Angelicin	trans-β-Farnesene	
Psoralen	α-p-Dimethylstyrene	
	Caryophyllene	
	α-Palmitolactone	
Fatty Acid Esters[62,104]		
n-Butylbutyrate	Phenylpropenes[62,104]	
n-Hexylbutyrate		
n-Octenyl-1-acetate		
n-Octanol		
n-Octyl-1-butyrate		
n-Octenyl-1-butyrate		
n-Decylacetate		
n-Octyl-1-caproate		

of secondary compounds (Table 1). Yet interactions among co-occurring secondary compounds as regards herbivory, insect and otherwise, have been largely ignored by investigators who have instead concentrated on single classes of secondary products. This emphasis is the result of both practical and theoretical considerations. Operationally, examining a single class of compounds facilitates the design and execution of experiments. Conceptually, stepwise insect/plant coevolution, as formalized by Ehrlich and Raven,[2] derived largely from experimental work on the Cruciferae;[3-5] the dramatic effects of mustard oil glycosides as toxins to nonadapted species[6,7] and as attractants to adapted species[8-12] led to the tacit assumption (and fervent hope) that single chemicals can provide the key to understanding insect/plant relationships.

Such, however, has seldom proved to be the case. Increasing evidence from behavioral and physiological studies of herbivorous insects suggests that interactions among secondary chemicals may well be the principal determinants underlying insect/plant coevolution. In the context of host recognition, insects that respond behaviorally and physiologically to "sign" or "token" stimuli, as do many crucifer-feeders to the presence of sinigrin,[12] exist but have proved rare.[13,14] The majority of insects studied instead appear to display a "response spectrum," whereby host acceptability depends..."not on the presence or absence of a single stimulant or deterrent but upon the total sensory impression derived from integrated response to multiple plant components".[15]

It is not altogether unexpected that, in a plant family as chemically complex as the Umbelliferae, most insect associates orient and respond to chemical mixtures. In fact, Dethier[15] developed his response spectrum hypothesis based partly on work with Papilio polyxenes, (Lepidoptera: Papilionidae) the umbellifer-feeding black swallowtail caterpillar, and its electrophysiological response to host and nonhost plant sap. Not only does the caterpillar respond to mixtures of chemicals, but ovipositing females, too, may respond preferentially to such mixtures. Attempts to characterize a single oviposition stimulant, despite intensive efforts, have been unsuccessful, presumably for this reason.[16] Yet another umbellifer associate, the root maggot Psila rosae (Diptera: Psilidae), orients both in larval and adult stages preferentially to mixtures of

compounds. This response is both behavioral and physiological. In field tests of orientation, adults flew to traps
baited with a mixture of trans-asarone and hexanal, two
biosynthetically distinct components of the essential oil
of Daucus carota (carrot), in greater numbers than to traps
with only trans-asarone, and air samples from the trans-
asarone-hexanal mixture evoked an electroantennogram response
greater than did either component singly.[17]

The Umbelliferae are by no means unique in this regard;
mixtures prevail as signals to associates of Solanaceae,[18]
Rosaceae,[19,20] Moraceae,[21] and Malvaceae,[22] to name a few.
There is even considerable evidence[9,23-25] that the crucifer-
sinigrin story may not be quite as simple as it once was
thought to be. Yet, while the importance of chemical mixtures in hostplant evaluation by insects has received some
measure of attention, the other side of the proverbial coin,
the importance of chemical mixtures in plant defense, by
comparison has received little. The virtual universality
of within-plant chemical diversity leads almost inescapably
to an adaptationist interpretation. McKey[26] has argued that
selection favors several lines of defense for several
reasons. Among these is the fact that chemicals may interact
either additively—where the total effective defense is equal
to the sume of the parts—or synergistically—where the total
effective defense is greater than the sum of the parts. This
can reduce the overall "cost of production"[26] of the defensive armamentarium. At the same time, plants can benefit
from chemical diversity in that exposing a herbivore to a
combination of chemicals can be expected to greatly delay
the acquisition of resistance.[27,28]

Resistance is "the developed ability in a strain [of
insects] to tolerate doses of toxicants that would prove
lethal to the majority of individuals in a normal population
of the same species".[29] It is generally a "preadaptive"
phenomenon; the toxic principle acts as a selective agent
removing susceptible genotypes so that resistant genotypes
already in the population increase in relative abundance.[30]
The extraordinary facility of insects to develop resistance
to synthetic organic insecticides has been the bugbear of
toxicological research; over 400 species of insects are known
to have developed resistance and over 60% of them are plant-
feeding agricultural pests[31] (Table 2). "Resistance
management" is a relatively new concept in which toxicologists devise chemical and operational techniques for

Table 2.   Insecticide resistance in agricultural pests.[31]

Insect Order	Number of Resistant Species	% Total Resistant Species (428)
Acarina	38	8.9
Coleoptera	64	14.9
Dermaptera	1	0.2
Heteroptera	16	3.7
Homoptera	42	9.8
Hymenoptera	3	0.7
Lepidoptera	64	14.9
Thysanoptera	7	1.6
Diptera	23	5.4
Total	258	60.1

prolonging the useful lifetime of a novel control chemical; among these techniques is the application of combinations of structurally related or unrelated insecticides.[32,33]   Plants, however, have long faced the problems of developed resistance in introducing novel control chemicals; resistance is in fact one of the driving forces in the Ehrlich and Raven[2] scenario. The production of several lines of chemical defenses, especially chemicals that can interact synergistically, may represent a plant's effort at prolonging the evolutionary viability of a particular chemical defense.

MECHANISMS OF RESISTANCE

     Resistance is a complex phenomenon since the toxicant, acting as the selective agent, can encounter obstacles preventing it from reaching its target site at many levels.[34] In order for an organism to suffer mortality from a toxicant, it must first come in contact with the latter; once contact is established, the toxicant must penetrate various physiological barriers to reach its target site; it must avoid metabolic alteration en route to the target site; and once at the target site it must compete for a position at the site and accommodate to changes in target site structure.

Behavioral Resistance

    Behavioral resistance results from the failure of the
insect to ingest or contact sufficient quantities of a toxin
to acquire a lethal dose.[30,35] Irritability, for example,
the decreased tendency of an insect to remain in contact with
a treated substrate is the basis for resistance to several
contact insecticides.[35] Feeding deterrency is another means
of behavioral resistance—insects that can detect the
presence of a chemical in their diet can respond by reducing
consumption rate or refusing to eat altogether,[36,37] thereby
failing to ingest a toxic dose. Finally, selective feeding
is also a means of behavioral resistance. Aphids, for
example, by feeding selectively on phloem sap, can feed with
aplomb on tobacco (Nicotiana tabacum) and other nicotine-
containing plants because they never come into contact with
the toxic agent.[38]

Physiological Resistance

    Physiological resistance involves the failure of an
allelochemical to reach its target site.[34,39] Reduced
uptake and transport can occur at many levels. At least
three factors influence the rate at which a xenobiotic can
penetrate insect cuticle—its lipid solubility in the highly
lipophilic epicuticle, its affinity for non-lipid cuticle
components such as protein and chitin, and its relative
solubility in hemolymph.[40] The cuticle is not the only
barrier to penetration by a foreign substance; midgut penetration can be reduced by alteration in the chemical
composition of the epithelial cell lining.[41] Alterations
in gut or blood pH, which can alter the ionic properties of
an ingested substance, can also confer resistance, since
ions cannot traverse the blood-brain barrier. Although over
98% of the nicotine in the blood of Manduca sexta (tobacco
hornworm) (Lepidoptera: Sphingidae) is present in the form
of the nicotinium ion, only the 2% remaining as a free base
is toxic. M. sexta is relatively immune to the effects of
nicotine by virtue of the fact that it can excrete the free
base rapidly, before it can reach the brain.[42] Physical
changes in the target site itself (as in "knock-down resistance" to pyrethroids in house flies) can also prevent a
toxin from causing mortality and thus can confer
resistance.[43]

Biochemical Resistance

Most known resistance to insecticidal agents, however, is biochemical in nature and involves enzymatic changes.[44,45] In many species of herbivorous insects, a suite of enzymes, generally associated with the microsomal fraction of cells of the gut wall, functions in converting lipophilic substances into hydrophilic substances for solubilization and excretion. This transformation from fat- to water-soluble more often than not results in detoxification; lipophilic xenobiotics present the greatest hazard to living organisms because tissues, organs and cells contain essential lipid-rich regions, such as membranes, and are therefore susceptible to attack.[46] One group of enzymes that effect these conversions are known collectively as mixed function oxidases (MFOs); they are capable of a number of metabolic conversions on a wide variety of substrates. Therefore, they are of central importance in the acquisition of resistance, both to synthetic organic insecticides and to plant allelochemicals.[44-47] With their tremendous biochemical flexibility, MFOs are ideally suited to counter the biochemical diversity of plants; rarely if ever is resistance attributable to a single, highly specific detoxifying enzyme (see Brattsten[44] for a discussion of rhodanese, an enzyme which converts toxic cyanide to the relatively nontoxic thiocyanate).

Resistance involving MFOs generally arises via selection of insect strains with high MFO activity. This may be the result of an aberrant form of a structural gene that determines the activity of an enzyme.[29,48] Resistance can also accrue from changes in a regulatory gene that controls the amount of an enzyme produced. Many of the MFOs are inducible —the presence in the gut of toxins initiates the de novo synthesis of MFO enzymes.[44-47] The extent to which enzymes are inducible varies widely among and within species, as does the opinion of investigators on the precise role of inducibility in resistance acquisition.[44,49,50]

MANAGEMENT OF RESISTANCE

By virtue of both empirical testing and hindsight, toxicologists have developed a few key rules in postponing the seemingly inevitable acquisition of resistance to a new pesticide. Insofar as resistance is an evolutionary

phenomenon, that is, selective removal by the toxic agent
of susceptible genotypes, it can be managed by deliberately
maintaining susceptible genotypes in the population. One
rather straightforward method to attain this goal is simply
to effect less than 100% mortality.[51] While this is an
innovation of sorts in pesticide technology, it is more or
less the status quo in plant defensive chemistry. Only a
very small minority of plant products have an acutely toxic
effect on generalized feeders upon contact or ingestion;
these constitute the conspicuous minority exploited commercially for control purposes by humans,[52] and, even in these
cases "the actual insecticidal action of plant extractives
may be due primarily to an artificially high level of
application, while, in fact, the parent plants are only
repellent in the field".[53]

Behavioral Resistance

        In contrast to acute toxicity, the most immediate
effect of the majority of known plant products on insects
is behavioral; most bring about a rapid reduction in
consumption rate. This phenomenon has been documented in
a sizable number of species for a startling number of plant
chemicals including aristolochic acids, cyclitols, chromenes, cyclopropenoid acids, phenolics, furanocoumarins,
polyacetylene—even alkaloids, reputed[54] to be among the
most acutely toxic plant products. With the number of
chemical compounds deterrent at ecological concentrations
even to exceedingly polyphagous noctuid caterpillars (e.g.,
Heliothis and Spodoptera spp.) (Table 3), it is indeed
remarkable that they find as many acceptable hostplants
as they do. The refusal of insects to continue to feed
in the presence of a toxicant with antifeedant properties
under bioassay no-choice experimental conditions likely
corresponds to movement away from the offending plant onto
a more suitable host in nature. Field trials confirm the
fact that insects, given a choice, move off plants with
antifeedants.[55] From the plant perspective, 100% kill is
not necessary—vacating the premises will suffice. If the
insect does not ingest the plant tissue in sufficient
quantity to acquire a toxic dose, it is then behaviorally
resistant to the toxins contained therein. As long as
opportunities are available in nature to switch hosts,
selection for physiological or biochemical resistance is
substantially reduced. Given the high species diversity
of herbaceous plant associations in which these plant

Table 3. Plant chemicals with deterrent properties to generalized herbivores (Lepidoptera: Noctuidae).[108-113]

Alkaloids	Lignans
Aristolochic acids	Monoterpenoids
Cardenolides	Non-protein amino acids
Chromenes	Phenolics
Cyclopropenoid acids	Phytoecdysones
Cyclitols	Polyacetylenes
Diterpenes	Saponins
Flavonoids	Sesquiterpenoids
Furanocoumarins	Tannins
Iridoid glycosides	Triterpenoids

products are often found (c.f. Feeny's[56] "unapparent plants"), such opportunities are undoubtedly available to generalist feeders.

Adams and Bernays[57] have demonstrated that mixtures of plant chemicals are more deterrent than single compounds. The reduction in consumption of bioassay filter discs treated with test chemicals and presented to Locusta migratoria (Orthoptera: Acrididae) was additive, that is, the reduction brought about by the mixture did not exceed the expected reduction calculated by summing the reduction brought about by the individual components of the mixture. However, even when each individual contribution of a test chemical to reduction in consumption was undetectable by bioassay, the collective effect was deterrent. Adams and Bernays[57] suggested that, due to additive effects of chemicals in low concentration, "it may well be that it is a better and more flexible defense strategy to possess such mixtures...than to have an accumulation of one such chemical in larger amount". Essential oils, often complex mixtures of dozens or hundreds of terpenoid and other constituents in relatively low concentration, may function generally throughout the plant kingdom as an antifeedant front line of defense; the consistent antifeedant effects[57] displayed by monoterpenoids, major essential oil components, lends at least some credence to this suggestion. Schoonhoven[58] and Jermy[59] agree that the "inhibitory biochemical profile" of a plant species is in all likelihood composed of several plant constituents and varies from species to species.

Physiological Resistance

Combinations of chemicals may also circumvent physiological resistance. Any agent that acts to enhance
penetration, transport or accessibility is referred to as a
quasi-synergist[39] since there is no real increase in the
lethality of the toxic agent, only an increase in the
efficiency of delivery to the target site. Any sort of
co-occurring chemical that enhances absorption of an allelochemical from the gut would aid in allowing a toxicant to
reach the still biochemically susceptible active site.
Saponins are known to act in this manner in mammalian
herbivores[26] and may do so in insects as well. They are
widespread among plants, occurring in over 500 species in
80 families,[60] and their presence in foliage is associated
with reduced suitability to a variety of herbivores.
Saponins can also displace endogenous lipids from lipoproteins[60] and may disrupt transport of xenobiotics by carrier
lipoporteins to insensitive storage areas such as fat body.
Finally, due to their hemolytic properties,[60] saponins may
facilitate the entry of toxicants into the cell to reach
the nucleus or other target site organelles by promoting
the breakdown in cell membrane integrity.

Systems other than those containing saponins may be
quasi-synergistic in nature. Oils are frequently added to
insecticide preparations to enhance penetration of the lipophilic cuticle.[40] Many plant products are sequestered in
oil-rich organs. In parsnip seeds, for example, furanocoumarins are localized in glandular structures in the seed
coat.[61] These structures, called vittae, also contain the
essential oils, rich in fatty acid esters.[62] By co-occurring
with such lipophilic materials, penetration by the marginally
polar furanocoumarins may be facilitated. Phototoxicity of
these compounds (vide infra) is greatly enhanced after
contact in the presence of a solvent; essential oils may act
as carrier solvents for the plant.

Management by Saturation

Georghiou[51] describes a number of techniques for resistance management that have parallels in angiosperm plants.
One means of countering biochemical resistance is "saturation"—flooding the system, as it were. Chief among
saturation techniques is the use of MFO inhibitors as
synergists. These are compounds that, by various mechanisms

interfere with the ability of MFOs to bind to substrates and to detoxify them. Commercially, 1,3-benzodioxole or methylenedioxyphenyl (MDP), compounds have been used with greatest success.[46] Ironically, it was with the use of these synergists that the role of MFOs in detoxification of pesticides was first elucidated.[63,64] MDPs are actually metabolized by MFOs and are believed to act as insecticide synergists by alternative substrate inhibition (competitive displacement at the active site) or by binding of the 1,3-benzodioxolium ion with the heme moiety of cytochrome P450.[65] By preventing the toxin from reaching the site of detoxification, synergists remove the selective advantage of alleles conferring resistance to the population, provided that alternate detoxification mechanisms do not exist or arise (Raffa, pers. comm.).

MDPs in particular, and synergists in general, reached scientific consciousness in a rather roundabout fashion. In 1940, several vegetable oils were compared as carrier substances for pyrethroid insecticide preparations; the pyrethroids displayed an unexpected increase in effectiveness when administered in sesame oil.[66] Subsequent isolation revealed that the agent responsible was sesamin, an MDP.[67] Continued research demonstrated that the 1,3-benzodioxole moiety was necessary for activity, and a number of synthetic analogues, including piperonyl butoxide, currently the most widely used of synergists,[46] were developed (Fig. 1). It was not until 1969, almost 30 years later, that sesamin was isolated from Chrysanthemum cinerarifolium (Compositae), the same plant that produces the pyrethroid insecticides.[68] The co-occurrence of the synergist with the plant product it acts with led to the speculation[69] that the compounds perform the same function in nature, that is, enhance the toxicity of co-occurring toxicants. Many authors have remarked on the phenomenon,[26,54,70,71] but evidence still remains scanty.

There are, however, at least two documented cases in which natural synergists do enhance the toxicity of co-occurring toxicants. Myristicin, an MDP with insecticide synergist properties,[77] is widely distributed among the Umbelliferae[72] and co-occurs extensively with furanocoumarins, known insecticidal substances.[73-75] When myristicin is administered in artificial diets at dosages corresponding to levels naturally occurring in foliage, it shows no appreciable toxicity to Heliothis zea (corn earworm)

INSECTICIDE
SYNERGIST

NATURAL
PRODUCT

piperonyl butoxide

myristicin (Umbelliferae)

o-methoxybenzaldehyde

o-anisaldehyde (Lauraceae)

1,4(5) substituted imidazoles

theobromine (Sterculiaceae)

o-nitroanisole

aristolochic acid (Aristolochiaceae)

benzylthiocyanate

benzylthiocyanate (Cruciferae)

aryl 2-propynl ether

capillin (Compositae)

Fig. 1.   Comparison of structures of known commercial
synergists and plant allelochemicals.   Insecticides from
references [81,84,114]; allelochemicals from references
[62,80,83,115,116].

(Lepidoptera: Noctuidae), a broadly polyphagous[76] caterpillar
(Table 4).   When, however, myristicin is incorporated into
diets containing 0.5% xanthotoxin, mortality to neonate
larvae due to xanthotoxin increases up to fivefold (Table
4).   Unlike piperonyl butoxide (PB), a commercial synergist,

Table 4. Mortality (%) of first instar Heliothis zea on
artificial diets. Berenbaum and Neal (in preparation).

Xanthotoxin (% Wet Wt.)	Myristicin (% Wet Weight)			
	0	0.01	0.03	0.10
0.000	1.1	0.0	0.0	0.0
0.100	0.0	6.7	0.0	6.7
0.250	20.0	13.3	43.3	86.7
0.375	23.3	40.0	38.5	93.3
0.500	16.7	30.0	60.0	86.7
1.500	50.0	100.0	96.7	
2.000	86.7			
LC_{50}	0.96	0.57	0.38	0.19
(± 95% confidence)	(.79-1.17)	(.48-.67)	(.32-.46)	(.17-.20)

synergism by myristicin increases with increasing amounts
— that is, there is a dose-dependent synergism at the
concentration of xanthotoxin investigated. Moreover, myristicin at high concentrations is superior to PB as a synergist
(Fig. 2). Although PB was formulated to meet certain
application and persistence requirements of toxicologists
(e.g., solubility in freon propellant), myristicin may have
been "formulated" over evolutionary time to suit more
precisely the toxicological requirements of the plant
producing it; while this idea is indeed appealing, it
remains sheer speculation.

Work by Miyakado et al.[77] lends credence to the suggestion by ecologists[70] and the practical demonstration by
toxicologists[63] that certain classes of secondary compounds
can become synergists themselves by acquiring a methylenedioxyphenyl substituent. Two co-occurring amides in Piper
nigrum (black pepper), pellitorine and piperine (Fig. 3),
are themselves relatively nontoxic to adzuki bean weevils
(Callosobruchus chinensis) (Coleoptera: Curculionidae);
when they are administered together in a 1:1 ratio, relative toxicity increases substantially (Table 5). Of these,
piperine is characterized by a methylenedioxy substituent.

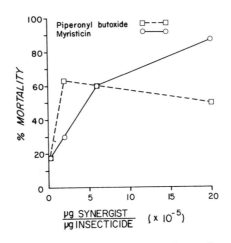

Fig. 2. Effect of varying concentration of two synergists
(piperonyl butoxide and myristicin) on dark toxicity of
xanthotoxin (0.5% wet weight of diet) to neonate Heliothis
zea (Lepidoptera: Noctuidae). (Berenbaum and Neal, in
preparation).

Pipericide (1)

Dihydropipercide (2)

Guineensine (3)

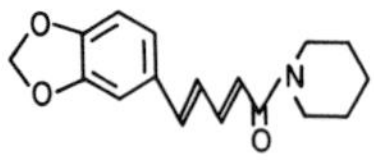

Piperine (4)

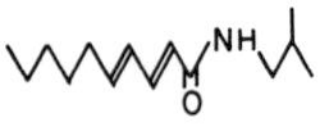

Pellitorine (5)

Piperstachine (6)

Fig. 3. Insecticidal amides from Piper nigrum (Piperaceae).[77]

Table 5. Insecticidal activity of black pepper compounds against male adults of Adzuki bean weevils (Callosobrochus chinensis L.)[77]

Compounds and Ratio	Observed Relative Toxicity (Pyrethrins = 1.00)
Pipercide (1)	0.23
Dihydropipercide (2)	0.56
Guineensine (3)	0.35
(1) + (2) + (3) = 1:1:1	1.15
(1) + (2) = 1:1	0.63
(1) + (3) = 1:1	0.70
(2) + (3) = 1:1	0.89
(1) + (2) + (3) = 75:5:20	0.80
Pellitorine (5)	0.02
Piperine (4)	<0.001
(5) + (4) = 1:1	0.23

Table 6. Plants producing compounds with methylenedioxy-phenyl rings.[78]

Amaryllidaceae	Menispermaceae
Annonaceae	Myristicaceae
Berberidaceae	Papaveraceae
Cactaceae	Piperaceae
Compositae	Orchidaceae
Fumariaceae	Ranunculaceae
Lauraceae	Rosaceae
Leguminosae	Rutaceae
Magnoliaceae	Umbelliferae

Three remaining amides—pipercide, dihydropipercide and guineensine, all with MDP substituents—are at least an order of magnitude more toxic than the unsubstituted pellitorine.

Methylenedioxyphenyls are widespread among plants—over 300 have been identified from 18 plant families, arising from a variety of biosynthetic pathways[78,79] (Table 6).

Although they have received much attention from the point of
view of synergism, the MDPs are in all probability only one
of several groups of natural synergists produced by plants.
While many natural products have not been tested for synergistic potential, they nonetheless (Fig. 1) share structural
similarities with known insecticide synergists. Organothio-
cyanates such as benzylthiocyanate in seeds of Lepidium
sativum (Cruciferae)[80] are active synergists of carbaryl,
pyrethrins and other insecticides.[65] Aryl propynl ethers,
commercial synergists, owe their toxicity to the presence
of acetylenic triple bonds;[65] these functional groups occur
commonly in polyacetylenes in the Umbelliferae[82] and
Compositae.[83] Acetylenic bonds are thought to form coordination monomers or polymers with metal ions in the microsomal
electron transport chain.[65] A number of other synergists
work by attacking metal ions and interfering with electron
transport.[84] O-Methoxybenzaldehyde, o-nitroanisole and
N-nitroso-N-phenyl-O-methylhydroxylamine are all thought to
act as blocked chelating agents in which mixed-function
oxidation unblocks the chelating agent to activate the
synergist. Naturally occurring analogues could conceivably
function in the same manner (Fig. 1).

    The search for synergists is hampered by the fact that
highly active synergists may themselves lack toxicity;
strictly speaking, the definition of "synergist" by conven-
tion implies a lack of inherent toxicity.[63] Thus myristicin
is a true synergist in view of the fact that it is not toxic
to Heliothis zea at synergistic dosages. Traditional
bioassay design, in which a single plant constituent is
incorporated into a completely defined or semisynthetic
artificial diet, would preclude the detection of synergistic
interactions.[58,59,85] On the other hand, given the enormous
chemical complexity of most plants (c.f., Table 1), it is
difficult a priori to select pairs (or even triads and
tetrads) of compounds to bioassay. One cannot even identify
synergists on the basis of concentration in the plant.
While they may be active in low concentrations, they do not
necessarily occur in low concentrations; for example, myristicin is the major constituent of parsnip fruit essential
oil.[62] Perhaps a more profitable approach would be to
search for natural products possessing functional groups in
common with known commercial synergists (e.g., Fig. 1) and
bioassaying these in combination with co-occurring compounds.
As for searching for new synergists, screening partitioned
extracts against purified compounds as a check would allow

the investigator to determine if the observed biological
activity of a particular plant is entirely attributable to
a single compound or if it is dependent at least in part
on the presence of other plant constituents.

Multiple Attack—Analogue Synergism

Multiple attack is another option in resistance
management[51]—that is administering a mixture of chemicals
with different toxicological mechanisms. Due to additive
effects, each constituent in a mixture can occur with
activity sufficiently low to avoid selection for resistance.
Structural modification of related chemicals may produce
compounds with markedly different toxicological properties;
this is the underlying rationale for observed "analog
synergism".[64] With few exceptions, plants produce not a
single compound of a particular chemical type but rather a
mixture of compounds sharing a common biosynthetic pathway.
This is certainly the case for terpenes,[62,86] glucosinolates,[25] cardenolides,[87] furanocoumarins,[88] alkaloids,[89]
cucurbitacins,[14] and tannins,[90] to name just a few. In
mixtures such as these, the potential for analogue synergism
certainly exists.

The mechanism underlying analogue synergism depends
upon the analogue series in question. In the case of
furanocoumarins, the primary mode of action is photobinding
via cyclo-adduct formation of furanocoumarins to DNA;
manifestations of UV-light induced effects include phototoxicity to bacteria, fungi, nematodes, insects, and
light-skinned birds and mammals.[73,91] Phototoxicity is
inherent in the psoralen ring system; the unsubstituted
psoralen is among the most phototoxic furanocoumarins.
Substitution that alters the electronic configuration[92] or
the shape of the molecule reduces or eliminates the phototoxicity.[91] Thus, xanthotoxin and bergapten (Fig. 4),
methoxylated in only a single position, are relatively
phototoxic; isopimpinellin (5,8-dimethoxypsoralen) is not.[93]
In the angular configuration, the double bond of the furan
ring is unable to form cyclo-adducts with pyrimidine bases
so, unlike the linear furanocoumarins, angular furanocoumarins are unable to crosslink DNA[91] and are thus relatively
nonphototoxic. The light-catalyzed crosslinkages enhances
effects on both DNA replication and template efficiency over
monoadduct formation.[93] There are, however, toxic properties of furanocoumarins that are independent of ultraviolet

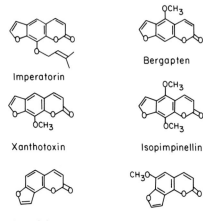

Imperatorin

Bergapten

Xanthotoxin

Isopimpinellin

Angelicin

Sphondin

Fig. 4.   Furanocoumarins in seeds and leaves of Pastinaca
sativa (Umbelliferae).

light; structural requirements for dark toxicity are not
nearly as well understood.[88]

     In Pastinaca sativa (parsnip) seeds there are at least
six different furanocoumarins[94] (Fig. 4). Of the six,
angelicin and sphondin are angular furanocoumarins and are
thus essentially nonphototoxic.[91] Of the remaining four
linear furanocoumarins, xanthotoxin and bergapten are
phototoxic, and isopimpinellin and imperatorin are not.[95]
When the furanocoumarin fraction from green parsnip fruits
is removed intact and incorporated into an artificial diet,
it displays greater toxicity to neonate Heliothis zea than
does an equimolar amount of pure xanthotoxin, the most
highly phototoxic furanocoumarin in the mixture. This
difference is most pronounced in the absence of ultraviolet
light (Fig. 5) and declines accordingly with increasing UV
(Berenbaum et al., in preparation). There is, then, toxicity associated with nonphotosensitizing furanocoumarins
that presumably operates via a completely different mechanism. When UV light is limiting, as it would be in wild
parsnips growing in shady habitats, dark-toxicity associated
with the nonphotosensitizing furanocoumarins enhances the
insecticidal effect. That this "multiple attack" prevails
in nature is suggested by the observation (Zangerl and

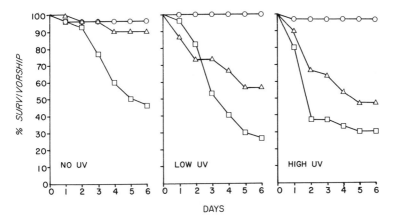

DAYS

Fig. 5. Comparison of toxicity of furanocoumarins alone
and in combination under various light regimes on first-
instar mortality and development time in Heliothis zea
(Lepidoptera: Noctuidae).
Open circles = control diet (no furanocoumarins), open
triangles = 0.25% xanthotoxin (wet weight of diet), open
squares = 0.20% furanocoumarins extracted intact from
immature seeds of Pastinaca sativa. (Composition of the
furanocoumarin mixture = 0.5% angelicin, 31.8% imperatorin,
18.1% bergapten, isopimpinellin 5.6%, xanthotoxin 35.5%,
sphondin 8.5%). 3-Way ANOVA among furanocoumarin treatment,
UV light level and day of death is significant (p = .018).
(Berenbaum et al., in preparation).

Berenbaum, in preparation) that shading plants significantly
increases the proportion of nonphototoxic furanocoumarins in
the foliage. A 20% reduction in incoming light increases
the proportion of nonphototoxic furanocoumarins (angelicin,
sphondin, imperatorin and isopimpinellin) from 43.6% to
62.2% (p < .001, t-test) (Zangerl and Berenbaum, in preparation), thus presumably maintaining the efficacy of the
mixture.

    That the furanocoumarins are differentially effective
as feeding deterrents may be responsible for the enhanced
toxicity of the furanocoumarin mixture. Yajima et al.[96]
bioassayed various furanocoumarins against Spodoptera
litura (Lepidoptera: Noctuidae), a polyphagous feeder, and
found over an order of magnitude difference in threshold

concentration of feeding inhibition. Curiously, the most
inhibitory furanocoumarin, the nonphototoxic isopimpinellin,
was effective at concentrations 1/20th that causing 50%
feeding inhibition by xanthotoxin, one of the most highly
phototoxic natural coumarins.

The phenomenon of dark vs light mediated toxicity
within a single class of secondary compounds is by no means
restricted to the furanocoumarins. Aromatic polyacetylenes
and thiophenes from the Compositae also have photodynamic
and nonphotodynamic mechanisms of toxicity. As with the
furanocoumarins, dark-related effects appear to involve
feeding inhibition and interference in nutrient metabolism.[97]
The great variety of polyacetylenes and thiophenes produced
within composite species may well result from synergistic or
additive interactions among the compounds due to their
different modes of action.

Multiple Attack—"Multichemical Defense"

Multiple attack may result not only from interactions
within related plant chemicals but also from the simultaneous
occurrence of biosynthetically unrelated chemicals. As
Schoonhoven[58] flatly states, "Plants never defend themselves
from insects with a monocomponent system". While the phenomenon is no doubt common, documentation is exceedingly rare.
One notable exception is the work by Kubo et al.[98] (see p.
171) on insect resistance in several species of trees. In
Podocarpus gracilior, for example, four norditerpenes are
both toxic and deterrent, two biflavones lack toxicity but
inhibit growth, and one phytoecdysteroid interferes with
molting and development; collectively these constitute a
"multichemical defense" against a variety of potential
herbivores.

CONCLUSIONS

When an insect bites into a leaf, it necessarily
encounters a mixture of chemicals[99] and its response,
behaviorally, ecologically, and evolutionarily, is to that
mixture. Complex mixtures are a common theme in insect
ecology and evolution—sex pheromones[100] and defensive
compounds,[101] for example, consist primarily of mixtures
of several different compounds. When presented with mixtures
of synthetic organic insecticides, many insects have devel-

oped resistance to those mixtures.[29] There is no reason, then, to suppose that naturally occurring toxicants cannot jointly subject insects to the same sort of selective pressures. By the same token, just as selection pressures from combinations of insecticides produce genetic linkage leading to cross-resistance or multiple resistance,[29] selective pressure exerted by herbivorous insects can produce genetic linkages among biochemically distinct groups of secondary chemicals, since they are jointly subject to the action of the selective agent.

Such a "whole-plant" response to selection may well account for two observations not readily explicable in the context of plant defense. First, many plant secondary compounds are present in trace (or smaller) amounts, well below any "insecticide dose" if indeed they are toxic at all. Their presence in the plant may result from selection for a synergistic interaction with a co-occurring toxicant, rather than selection for insecticidal properties per se. Secondly, genetic linkage in the biosynthesis and distribution of distinct groups of allelochemicals may account for the seasonal, diurnal or even hourly variation in the levels of production of many classes of secondary chemicals.[102] While the variation itself may not appear "adaptive" or "defensive", it may actually represent enhancement of toxicity in the context of simultaneous variation in other plant tissue components.

It is a truism to assert that different plant parts cannot evolve independently of one another—to state, for example, root evolution proceeds entirely without influence of stem, leaf or seed is to assert that these plant parts are distinct populations with separate gene pools. Nonetheless, people seem willing to accept the idea that secondary compounds within a plant can evolve independently of other groups of plant secondary compounds. It is difficult to imagine how such a scenario could operate given that insects and other plant enemies respond to and act upon the entire plant phenotype. To understand the dynamics of the coevolution between insects and plants, it is necessary to understand the ways in which classes of secondary compounds within a plant themselves coevolve, that is, reciprocally act as selective agents upon each other. After all, it is the joint effect that provides the defense against enemies.

An analogy can be drawn with the familiar fairy tale by
the Brothers Grimm[103] relating the adventures of the Brementown musicians—an ass, a dog, a cat, and a cock who had
outlived their usefulness to their human owners and took to
the road in an effort to find a better life as musicians in
Bremen. On their way they passed a robber's house well
provisioned with

> "eatables and drinkables...They consulted together
> how it should be managed so as to get the robbers
> out of the house, and at last they hit on a plan.
> The ass was to place his forefeet on the window-
> sill, the dog was to get on the ass's back, the
> cat on top of the dog and lastly the cock was to
> fly up and perch on the cat's head. When that was
> done, at a given signal they all began to perform
> their music. The ass brayed, the dog barked, the
> cat mewed, and the cock crowed; then they burst
> through into the room, breaking all the panes of
> glass. The robbers fled at the dreadful sound;
> they thought it was some goblin and fled to the
> woods in the utmost terror. Then the four compan-
> ions sat down to table, made free with the remains
> of the meal, and feasted as if they had been
> hungry for a month. And when they had finished
> they put out the lights, and each sought out a
> sleeping-place to suit his nature and habits."[103]

Each to suit nature and habits, as each allelochemical
contributes to defense according to its chemical nature and
habits. With this principle in mind, prospects for understanding interactions among allelochemicals are anything
but Grimm.

ACKNOWLEDGMENTS

I thank Drs. Richard Larson, Robert Metcalf and David
Seigler of the University of Illinois at Urbana-Champaign
and John Andersen of the Northern Regional Research Center
in Peoria, Illinois for comments on the manuscript and I
thank my collaborators and friends, Ellen Heininger,
Jonathan Neal, James Nitao, Steve Sheppard, and Arthur
Zangerl, for last-minute assistance above and beyond the
call of duty in analyzing data, locating literature, drawing
figures, checking facts, proofreading, criticizing the

manuscript, and generally offering enthusiastic support and interest. Research funded by U.S.D.A. grant AG83 CRCR-1-1226.

REFERENCES

1. FEENY, P. 1977. Defensive ecology of the Cruciferae. Ann. Missouri Bot. Gard. 64: 221-234.
2. EHRLICH, P.R., P.H. RAVEN. 1964. Butterflies and plants: a study in coevolution. Evolution 18: 586-608.
3. VERSCHAFFELT, E. 1911. The case determining the selection of food in some herbivorous insects. Proc. Acad. Sci. Amsterdam 13: 536-542.
4. THORSTEINSON, A.J. 1953. The chemotactic responses that determine host specificity in an oligophagous insect (Plutella maculipennis (Curt.) Lepidoptera). Canad. J. Zool. 31: 52-72.
5. THORSTEINSON, A.J. 1960. Host selection in phytophagous insects. Ann. Rev. Entomol. 5: 193-218.
6. ERICKSON, J.M., P. FEENY. 1974. Sinigrin: a chemical barrier to the black swallowtail butterfly, Papilio polyxenes. Ecology 55: 103-111.
7. BLAU, P.A., P. FEENY, L. CONTARDO, D.S. ROBSON. 1978. Allylglucosinolate and herbivorous caterpillars: a contrast in toxicity and tolerance. Science 200: 1296-1298.
8. DAVID, W.A., B.O.C. GARDINER. 1962. Oviposition and the hatching of the eggs of Pieris brassicae (L.) in a laboratory culture. Bull. Entomol. Res. 53: 91-109.
9. FEENY, P., K.L. PAAUWE, N.J. DEMONG. 1970. Flea beetles and mustard oils: host plant specificity of Phyllotreta cruciferae and P. striolata adults (Coleoptera: Chrysomelidae). Ann. Entomol. Soc. Am. 63: 832-841.
10. HICKS, K.L. 1974. Mustard oil glycosides: feeding stimulants for adult cabbage flea beetles, Phyllotreta cruciferae (Coleoptera: Chrysomelidae). Ann. Entomol. Soc. Am. 67: 261-264.
11. HAWKES, C., T.H. COAKER. 1979. Factors affecting the behavioral responses of the adult cabbage root fly, Delia brassicae, to host plant odor. Entomol. Exp. Appl. 25: 45-58.

12. SCHOONHOVEN, L.M. 1967. Chemoreception of mustard
    oil glucosides in larvae of Pieris brassicae.
    Proc. Kon. Ned. Akad. Wtsch. C. 70: 556-563.
13. REES, J.C. 1969. Chemoreceptor specificity associated
    with choice of feeding site by the beetle,
    Chrysolina brunsvicensis, on its foodplant,
    Hypericum hirsutum. Entomol. Exp. Appl. 12:
    565-583.
14. METCALF, R.L., A.M. RHODES, R.A. METCALF, J. FERGUSON,
    E.R. METCALF, P-Y. LU. 1982. Cucurbitacin
    contents and diabroticite (Coleoptera: Chrysome-
    lidae) feeding upon Cucurbita spp. Environ.
    Entomol. 11: 931-937.
15. DETHIER, V.G. 1973. Electrophysiological studies of
    gustation in lepidopterous larvae. II. Taste
    spectra in relation to food-plant discrimination.
    J. Comp. Physiol. 82: 103-134.
16. FEENY, P., L. ROSENBERRY, M. CARTER. 1983. Chemical
    aspects of oviposition behavior in butterflies.
    In Herbivorous Insects: Host-seeking Behavior and
    Mechanisms. (S. Ahmad, ed.), Academic Press, New
    York, pp. 27-76.
17. GUERIN, P.M., E. STÄDLER, H.R. BUSER. 1983. Identi-
    fication of host plant attractants for the carrot
    fly, Psila rosae. J. Chem. Ecol. 9: 843-861.
18. MAY, M.L., S. AHMAD. 1983. Host location in the
    Colorado potato beetle: searching mechanisms in
    relation to oligophagy. In S. Ahmad, ed., op. cit.
    Reference 16, pp. 173-199.
19. RODRIGUEZ, J.G., T.R. KEMP, Z.T. DABROWSKI. 1976.
    Behavior of Tetranychus urticae toward essential oil
    mixtures from strawberry foliage. J. Chem. Ecol.
    2: 221-230.
20. LADD, T.L., T.P. McGOVERN. 1980. Japanese beetle:
    enhancement of lures by eugenol and caproic acid.
    J. Econ. Entomol. 73: 718-720.
21. WATANABE, T. 1958. Substances in mulberry leaves
    which attract silkworm (Bombyx mori). Nature 182:
    325-326.
22. HEDIN, P.A., A.C. THOMPSON, R.C. GUELDNER. 1976.
    Cotton plant and insect constituents that control
    boll weevil behavior and development. Rec. Adv.
    Phytochem. 10: 271-350.
23. FINCH, S. 1978. Volatile plant chemicals and their
    effect on host finding by the cabbage root fly (Delia
    brassicae). Entomol. Exp. Appl. 24: 350-359.

24. NIELSEN, J.K. 1978. Host plant discrimination within Cruciferae: feeding responses of four leaf beetles to glucosinolates, cucurbitacins and cardenolides. Entomol. Exp. Appl. 24: 41-54.

25. RODMAN, J.E., F.S. CHEW. 1980. Phytochemical correlates of herbivory in a community of native and naturalized Cruciferae. Biochem. Syst. Ecol. 8: 43-50.

26. McKEY, D. 1979. The distribution of secondary compounds within plants. In Herbivores, Their Interaction With Secondary Plant Metabolites. (G.A. Rosenthal, D.H. Janzen, eds.), Academic Press, New York, pp. 55-133.

27. PIMENTEL, D., A.C. BELLOTTI. 1976. Parasite-host population systems and genetic stability. Am. Nat. 110: 877-888.

28. FEENY, P. 1983. Coevolution of plants and insects. In Natural Products for Innovative Pest Management. (D.L. Whitehead, W.S. Bowers, eds.), Pergamon Press, Oxford, pp. 167-185.

29. OPPENOORTH, F.J., W. WELLING. 1976. Biochemistry and physiology of resistance. In Insecticide Biochemistry and Physiology. (C.F. Wilkinson, ed.), Plenum Press, New York, 768 pp.

30. BROWN, A.W., R. PAL. 1971. Insecticide Resistance in Arthropods. World Health Organization, Geneva, pp. 491.

31. GEORGHIOU, G.P., R.B. MELLON. 1983. Pesticide resistance in time and space. In Pest Resistance to Pesticides. (G.P. Georghiou, T. Saito, eds.), Plenum Press, New York, pp. 1-46.

32. GEORGHIOU, G.P. 1983. Management of resistance in arthropods. In G.P. Georghiou, T. Saito, eds., op. cit. Reference 31, pp. 769-792.

33. OZAKI, K. 1983. Suppression of resistance through synergistic combinations with emphasis on plant-hoppers and leafhoppers infesting rice in Japan. In G.P. Georghiou, T. Saito, eds., op. cit. Reference 31, pp. 595-614.

34. BROOKS, G.T. 1976. Penetration and distribution of insecticides. In C.F. Wilkinson, ed., op. cit. Reference 29, pp. 3-60.

35. PLUTHERO, F.G., R.S. SINGH. 1984. Insect behavioural responses to toxins: practical and evolutionary considerations. Can. Entomol. 116: 57-68.

36. CHAPMAN, R.F., E.A. BERNAYS. 1972. The chemical
     resistance of plants to insect attack. Pont. Acad.
     Sci. Scripta Var. 41: 603-643.
37. BERNAYS, E.A., R.F. CHAPMAN. 1977. Deterrent
     chemicals as a basis of oligophagy in Locusta
     migratoria (L.). Ecol. Entomol. 2: 1-18.
38. GUTHRIE, F.E., W.V. CAMPBELL, R.L. BARON. 1962.
     Feeding sites of the green peach aphid with respect
     to its adaptation to tobacco. Ann. Entomol. Soc.
     Am. 55: 42-46.
39. SUN, Y.P., E.R. JOHNSON. 1972. Quasi-synergism and
     penetration of insecticides. J. Econ. Entomol. 65:
     349-353.
40. BROOKS, G.T. 1976. Penetration and distribution of
     insecticides. In C.F. Wilkinson, ed., op. cit.
     Reference 29, pp. 3-58.
41. HOLLINGWORTH, R.M. 1976. The biochemical and physiol-
     ogical basis of selective toxicity. In C.F.
     Wilkinson, ed., op. cit. Reference 29, pp. 431-506.
42. SELF, L.S., F.E. GUTHRIE, E. HODGSON. 1964. Adapta-
     tion of tobacco hornworms to the ingestion of
     nicotine. J. Insect Physiol. 10: 907-914.
43. FARNHAM, A.W. 1977. Genetics of resistance of
     houseflies (Musca domestica L.) to pyrethroids.
     I. Knock-down resistance. Pest. Sci. 8: 631-636.
44. BRATTSTEN, L.B. 1979. Biochemical defense mechanisms
     in herbivores against plant allelochemicals. In
     G.A. Rosenthal, D.H. Janzen, eds., op. cit.
     Reference 26, pp. 199-271.
45. BRATTSTEN, L.B. 1979. Ecological significance of
     mixed function oxidations. Drug Metab. Rev. 10:
     35-58.
46. WILKINSON, C.F. 1976. Insecticide interactions. In
     C.F. Wilkinson, ed., op. cit. Reference 29, pp.
     605-648.
47. BRATTSTEN, L.B., C.F. WILKINSON, T. EISNER. 1977.
     Herbivore plant interactions: mixed-function
     oxidases and secondary plant substances. Science
     196: 1349-1352.
48. PLAPP, F.W., T.C. WANG. 1983. Genetic origins of
     insecticide resistance. In G.P. Georghiou, T.
     Saito, eds., op. cit. Reference 31, pp. 47-70.
49. WILKINSON, C.F. 1983. Role of mixed-function oxidases
     in insecticide resistance. In G.P. Georghiou, T.
     Saito, eds., op. cit. Reference 31, pp. 175-206.

50. TERRIERE, L.C. 1983. Enzyme induction, gene amplification and insect resistance to insecticides. In G.P. Georghiou, T. Saito, eds., op. cit. Reference 31, pp. 265-298.

51. GEORGHIOU, G.P. 1983. Management of resistance in arthropods. In G.P. Georghiou, T. Saito, eds., op. cit. Reference 31, pp. 769-792.

52. JACOBSON, M., D.G. CROSBY. 1971. Naturally Occurring Insecticides. Marcel Dekker, Inc., New York, 585 pp.

53. CROSBY, D.G. 1966. Natural pest control agents. In Natural Pest Control Agents. (R.F. Gould, ed.), Advances in Chemistry Series 53, American Chemical Society, Washington, pp. 1-16.

54. RHOADES, D.F., R.G. CATES. 1976. A general theory of plant antiherbivore chemistry. Rec. Adv. Phytochem. 10: 168-213.

55. BERNAYS, E.A. 1983. Antifeedants in crop pest management. In D.L. Whitehead, W.S. Bowers, eds., op. cit. Reference 28, pp. 259-271.

56. FEENY, P. 1976. Plant apparency and chemical defense. Rec. Adv. Phytochem. 10: 1-40.

57. ADAMS, C.M., E.A. BERNAYS. 1978. The effects of combinations of deterrents on the feeding behaviour of Locusta migratoria. Entomol. Exp. Appl. 23: 101-109.

58. SCHOONHOVEN, L.M. 1982. Biological aspects of antifeedants. Entomol. Exp. Appl. 31: 57-69.

59. JERMY, T. 1983. Multiplicity of insect antifeedants in plants. In D.L. Whitehead, W.S. Bowers, eds., op. cit. Reference 28, pp. 223-236.

60. APPELBAUM, S.W., Y. BIRK. 1979. Saponins. In G.A. Rosenthal, D.H. Janzen, eds., op. cit. Reference 26, pp. 539-566.

61. LADYGINA, E.U., V.A. MAKAROVA, N.S. IGNAT'EVA. 1970. Morphological and anatomical description of Pastinaca sativa fruit and localization of the furocoumarins in them. Farmatsiya 19: 29-35 (in Russian, CA 1970, 74: 61588x).

62. STAHL, E., K.H. KUBECZKA. 1979. Über atherische öle der Apiaceae (Umbelliferae). VI. Untersuchungen zum Vorkommen von Chemotypen bei Pastinaca sativa L. Planta Med. 37: 49-56.

63. METCALF, R.L. 1967. Mode of action of insecticide synergists. Ann. Rev. Entomol. 12: 229-256.

64. CASIDA, J.E. 1970. Mixed-function oxidase involvement in the biochemistry of insecticide synergists. J. Agr. Food Chem. 18: 753-760.
65. TAHORI, A.S. (ed.). 1971. Insecticide Resistance, Synergism, Enzyme Induction. Gordon and Breach, New York, 302 pp.
66. EAGLESON, C. 1942. Sesame oil as a synergist of pyrethrum insecticides. Soap and Sanit. Chem. 18: 125-127.
67. HALLER, H.L., E.R. McGOVRAN, L.D. GOODHUE, W.N. SULLIVAN. 1942. The synergistic action of sesamin with pyrethrum insecticides. J. Org. Chem. 7: 183-185.
68. DOSKOTCH, R.W., F.S. EL-FERALY. 1969. Isolation and characterization of (+) sesamin and β-cyclopyrethrosin from pyrethrum flowers. Can. J. Chem. 47: 1139-1142.
69. KRIEGER, R.I., P.P. FEENY, C.F. WILKINSON. 1971. Detoxication enzymes in the guts of caterpillars: an evolutionary answer to plant defenses? Science 172: 579-581.
70. MILLER, J.S., P. FEENY. 1983. Effects of benzyliso-quinoline aldaloids on the larvae of polyphagous Lepidoptera. Oecologia 58: 332-339.
71. SCRIBER, J.M. 1983. Host-plant suitability. In Chemical Ecology of Insects. (W.J. Bell, R.T. Carde, eds.), Sinauer Associates, Sunderland, Massachusetts, pp. 159-202.
72. HARBORNE, J.B., V.H. HEYWOOD, C.A. WILLIAMS. 1969. Distribution of myristicin in seeds of the Umbelliferae. Phytochem. 8: 1729-1732.
73. BERENBAUM, M. 1978. Toxicity of a furanocoumarin to armyworms: a case of biosynthetic escape from insect herbivores. Science 201: 532-534.
74. BERENBAUM, M. 1983. Coumarins and caterpillars: a case for coevolution. Evolution 37: 163-179.
75. BERENBAUM, M., P. FEENY. 1981. Toxicity of angular furanocoumarins to swallowtails: escalation in a coevolutionary arms race? Science 212: 927-929.
76. KOGAN, J., D.K. SELL, R.E. STINNER, J.R. BRADLEY, M. KOGAN. 1978. V. A Bibliography of Heliothis zea (Boddie) and H. virescens (F.) (Lepidoptera: Noctuidae). International Agricultural Publications INTSOY Series Number 17, Urbana, Illinois, 242 pp.
77. MIYAKADO, M., I. NAKAYAMA, N. OHNO, H. YOSHIOKA. 1983. Structure, chemistry and actions of the Piperaceae amids: new insecticidal constituents isolated from

the pepper plant. In Natural Products for Innovative Pest Management. (D.L. Whitehead, ed.), Pergamon Press, New York, pp. 369-382.
78. NEWMAN, A.A. 1962. The occurrence, genesis and chemistry of the phenolic methylenedioxy ring in nature. Chem. Prod. 25: 161-166.
79. LICHTENSTEIN, E.P., J.E. CASIDA. 1963. Myristicin, an insecticide and synergist occurring naturally in the edible parts of parsnip. J. Agric. Food Chem. 11: 410-415.
80. LÜTHY, J., M.H. BENN. 1977. Thiocyanate formation from glucosinolates: a study of the autolysis of allylglucosinolate in Thlaspi arvense L. seed flour extracts. Can. J. Biochem. 55: 1028-1031.
81. WILKINSON, C.F. 1971. Insecticide synergists and their mode of action. In A.S. Tahori, ed., op. cit. Reference 65, pp. 117-160.
82. BOHLMANN, F. 1971. Acetylenic compounds in the Umbelliferae. Bot. J. Linn. Soc. 64: Suppl. 1: 279-291.
83. BOHLMANN, F., W. SUCROW. 1963. Natürlich vorkommende Acetylenverbindungen. In Modern Methods of Plant Analysis, Vol. 6. (H.F. Linskens, M.V. Tracey, eds.), Springer-Verlag, Berlin, pp. 81-108.
84. HENNESEY, D.J. 1971. The design of synergists for large scale application. In A.S. Tahori, ed., op. cit. Reference 65, pp. 161-166.
85. BERENBAUM, M. 1985. Post-ingestive effects of allelochemicals on insects, on Paracelsus and plant products. In Insect-plant Interactions. (T.A. Miller, J. Miller, eds.), Springer-Verlag, New York. Chapter 5 (in press).
86. JOHNSON, A.E., H. NURSTEN, A. WILLIAMS. 1971. Vegetable volatiles: a survey of components identified: Part II. Chem. Ind. 1971: 1212-????.
87. BROWER, L.P., J.N. SEIBER, C.J. NELSON, S.P. LYNCH, P.M. TUSKES. 1982. Plant-determined variation in the cardenolide content, thin-layer chromatography profiles, and emetic potency of monarch butterflies, Danaus plexippus reared on the milkweed, Asclepias eriocarpa in California. J. Chem. Ecol. 8: 579-633.
88. MURRAY, R.D.H., J. MENDEZ, S.A. BROWN. 1982. The Natural Coumarins. John Wiley and Sons Ltd., Chichester.

89. WATERMAN, P.G. 1975. Alkaloids of the Rutaceae:
     their distribution and systematic significance.
     Biochem. Syst. Ecol. 3: 149-180.
90. ZUCKER, W.V. 1983. Tannins: does structure determine
     function? An ecological perspective. Am. Nat. 121:
     335-365.
91. SCOTT, B.R., M.A. PATHAK, G.R. MOHN. 1976. Molecular
     and genetic basis of furanocoumarin reactions.
     Mutat. Res. 39: 29-74.
92. PATHAK, M.A., L.R. WORDEN, K.D. KAUFMAN. 1967. Effect
     of structural alterations on the photosensitizing
     potency of furocoumarins (psoralens) and related
     compounds. J. Invest. Dermatol. 48: 103-110.
93. BROWN, S.A. 1979. Biochemistry of the coumarins.
     Rec. Adv. Phytochem. 12: 249-286.
94. BERENBAUM, M., A.R. ZANGERL, J.K. NITAO. 1984.
     Furanocoumarins in seeds of wild and cultivated
     parsnip (Pastinaca sativa). Phytochem. 23: 1809-1810.
95. MUSAJO, L., G. RODIGHIERO. 1962. The skin-
     photosensitizing furocoumarins. Experientia 18:
     153-200.
96. YAJIMA, T., N. KATO, K. MUNAKATA. 1977. Isolation
     of insect anti-feeding principles in Orixa japonica
     Thunb. Agric. Biol. Chem. 41: 1263-1268.
97. CHAMPAGNE, D.T., J.T. ARNASON, B.J.R. PHILOGENE, J.
     LAM. 1984. Phototoxic effects of natural poly-
     acetylenes and thiophenes on insect herbivores.
     Phytochem. Soc. No. Am. Newsletter 24: 28
     (abstract).
98. KUBO, I. 1984. Multichemical insect and fungal
     resistance in plants. Rec. Adv. Phytochem. 19:
     171-194.
99. JANZEN, D.H. 1973. Community structure of secondary
     compounds in nature. Pure Appl. Chem. 34: 529-538.
100. CARDE, R.T., T.C. BAKER. 1983. Sexual communication
     with pheromones. In W.J. Bell, R.T. Carde, eds.,
     op. cit. Reference 71, pp. 355-524.
101. BLUM, M. (ed.). 1981. Chemical Defenses of Arthropods.
     Academic Press, New York, 562 pp.
102. SEIGLER, D.S. 1977. Primary roles for secondary
     compounds. Biochem. Syst. Ecol. 5: 195-199.
103. CRANE, L. (trans.). 1963. Household Stories from the
     Collection of Brothers Grimm. Dover Publications,
     New York, 269 pp.
104. KUBECZKA, K.H., E. STAHL. 1975. Über atherische
     öle der Apiaceae (Umbelliferae). I. Das Würzelöl

von Pastinaca sativa. Planta Medica 27: 235-241.

105. HEGNAUER, R. 1964-1973. Chemotaxonomie der Pflanzen. 6 Volumes, Birkhauser Verlag, Basel.

106. STECK, W., B.K. BAILEY. 1969. Characterization of plant coumarins by combined gas chromatography, ultraviolet absorption spectroscopy and NMR analysis. Can. J. Chem. 47: 3577-3583.

107. IVIE, G.W., D.L. HOLT, M.C. IVEY. 1981. Natural toxicants in human foods; psoralens in raw and cooked parsnip root. Science 213: 909-910.

108. KLOCKE, J.A., B. CHAN. 1982. Effects of cotton condensed tannin on feeding and digestion in the cotton pest, Heliothis zea. J. Insect Physiol. 28: 911-915.

109. KOUL, O. 1982. Insect feeding deterrents in plants. Indian Rev. Life Sci. 2: 97-125.

110. KUBO, I., J. KLOCKE. 1983. Isolation of phytoecdysones as insect ecdysis inhibitors and feeding deterrents. In Plant Resistance to Insects. (P. Hedin, ed.), A.C.S. Symposium Series 208, American Chemical Society, Washington, D.C., pp. 329-346.

111. RAUSHER, M.D. 1979. Coevolution in a Simple Herbivore-Plant System. Doctoral dissertation, Cornell University, Ithaca, New York.

112. WAISS, A.C., B. CHAN, C. ELIGER, D. DREYER, R. BINDER, R. GUELDNER. 1981. Insect growth inhibitors in crop plants. Bull. Entomol. Soc. Am. 27: 217-221.

113. WISDOM, C., J.T. SMILEY, E. RODRIGUEZ. 1983. Toxicity and deterrency of sesquiterpene lactones and chromenes to the corn earworm (Lepidoptera: Noctuidae). J. Econ. Entomol. 76: 993-998.

114. WILKINSON, C.F. 1973. Insecticide synergism. Chemtech (August): 492-497.

115. GOODWIN, T.W., E.I. MERCER. 1972. Introduction to Plant Biochemistry. 1st Edition, Pergamon Press, New York.

116. KARRER, W. 1958. Konstitution und Vorkommen der Organischen Pflanzenstoffe. Birkhauser Verlag, Basel.

Chapter Seven

MULTIFACETED CHEMICALLY BASED RESISTANCE IN PLANTS

ISAO KUBO AND FREDERICK J. HANKE

Division of Entomology and Parasitology
College of Natural Resources
University of California
Berkeley, California 94720

INTRODUCTION

When one considers the fairly universal primary metabolism among plants, it might seem reasonable to expect them
to be relatively equal in susceptibility to bacterial, fungal
or viral pathogens as well as to animal attack. However, it
is quite obvious that is not the case. Everyone has seen
evidence of plant resistance to pathogens or animal attack
in their own garden. One plant is covered with a blanket of
aphids or blight while another species is relatively unharmed.

Some resistance of plants is based on physical methods,
such as the mimicry of Passiflora cyanea which forms dummy
eggs of the Heliconius butterfly on its leaves to discourage
actual Heliconius egg laying.[1] Another example of physical
resistance is found in cotton (Gossypium hirsutum) where
resistance to the leaf hopper Empoasca fabae has been shown
to be related to the hairiness of the leaves.[2] However, it
is generally believed that most defenses in plants are
chemical in origin.[3-5] Several plants are rather spectacular
in their ability to ward off attack through chemistry, even
after their death, and have been put to use by man. Odif-
erous cedar chests protect otherwise vulnerable textiles from
insect attack. Teak houses and furniture in southeast Asia

are subject to seemingly voracious tropical insects and
potentially damaging fungi yet remain intact for over a
hundred years. More direct applications of plant chemistry
are also readily apparent. The fruit of Azadiracta indica,
the Indian Neem tree, which was recently shown to contain
several powerful insect antifeedants,[6-8] has been used to
repel insects on crops for over a thousand years. The
commercial insecticide rotenone, in use for over fifty years,
was originally extracted from the roots of Derris elliptica.[9]

The amount of information about the chemically based
resistance of plants has increased enormously in the last
few years. This is primarily due to modern instrumentation
speeding up both the isolation and the identification of
compounds, and also to increased interest in finding new and
useful natural products. Nevertheless, there still are few
reports of single plant species containing multiple types of
chemically based resistances towards the variety of environmental hazards to which most plants are continually exposed.

Seldom is a plant, even when in a unique environment,
placed under stress by only one organism. It seems likely
that evolutionary pressures, suspected of giving rise to
offensive and defensive secondary metabolites,[10] would be
present for several different types of stress at the same
time. This could give rise to the simultaneous or sequential
evolution of secondary metabolites both diverse in structure
and/or activity. We believe that the lack of reported multifaceted chemical resistance reflects the methods and interests
of past researchers more than the actual chemical constituents
of plants, and that reports of this type will increase in the
near future. We have found multiple bioassay directed isolation to be most effective in probing the different types of
chemically based resistance potentially present in plants.
This insures that an emphasis is not placed on those compounds
which are easier to isolate or more abundant but rather upon
active compounds. These in-house bioassays elicit potential
molluscicidal properties, effects on monocotyledonous and
dicotyledonous seed germination and growth, insect toxicity
and insect antifeedant activity. Compounds are also tested
for their effect on bacteria and fungi. From investigations
utilizing these assays, we have observed examples of individual plant species containing a range of resistances relying
upon or at least augmented by various chemical components.
We would like to report here the identification of compounds
responsible for a multifaceted chemically based resistance in

species from three different families of plants. These are
the east African medicinal plant Ajuga remota (Labiatae),
the south African evergreen Podocarpus gracilior (Podocarpaceae) and the bitter olive Olea europaea (Oleaceae).

FAMILY LABIATAE

Ajuga remota is an east African medicinal mint species
of the family Labiatae. Both the bitter testing leaves and
roots are known to be resistant to insect attack, and a
survey of vegetation after an outbreak of locust on the
Kenyan savanna revealed that this was the only plant which
survived the assault. This observation led us to investigate
the possibility of a chemical defense being responsible. The
leaves and roots of A. remota were extracted with methanol.
After removal of solvent, the oil was partitioned between
water and hexane and the aqueous layer sequentially extracted
with ethyl ether and ethyl acetate. This gave four fractions
(hexane, ether, ethyl acetate and water-soluble fractions) of
increasing polarity. These four fractions were then tested
for biological activity. The ether-soluble portion tested
positive in the choice host plant leaf disk assay using
Spodoptera exempta. In this test (Fig. 1) young leaves
coated with a test substance and young unadulterated leaves
are presented to insects and avoidance of the coated leaves
is interpreted as a positive antifeedant effect. Subsequent
leaf disk bioassays directed the purification of three
active compounds which we named ajugarin-I (1), -II (2) and
-III (3).[11] Their structures were identified by spectral
means and by comparisons with the related known diterpene
clerodin (4) and its derivatives. Clerodin (4) itself was
later isolated as an antifeedant as well.[12] Antifeedant
activity is now recognized as being very important, and
these diterpenes are members of a pool of compounds which
may be used as a source to control insects without the
toxicity problems now recognized to be wide-spread.[13]

The choice host plant leaf disk assay is specific and
tests only for avoidance behavior to a substance. With a
choice available, the animal is not forced to eat the
substance and toxicity information could be overlooked.
Also, the test is short term so that any effects that would
take time to observe could be missed. We normally use
another assay to direct the isolation of toxic or growth
inhibition factors. The artificial diet feeding assay

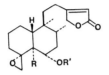

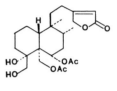

<u>1</u>, R= CH₂OAc, R'= Ac
<u>2</u>, R= CH₂OAc, R'= OH
<u>6</u>, R= CH₃, R'= Ac

<u>3</u>

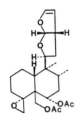

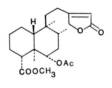

<u>4</u>

<u>5</u>

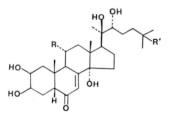

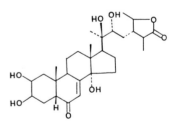

<u>7</u>, R= H, R'= OH
<u>9</u>, R= OH, R'= H
<u>20</u>, R= R'= H

<u>8</u>

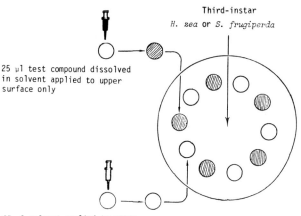

Fig. 1. The choice host plant leaf disk assay. Treated and untreated leaf disks from a host plant (e.g., young cotton leaves) are placed in an alternating pattern in foam grids inside a glass petri dish. Larvae are placed inside the dish and allowed to feed for 48 h (25°, 80% relative humidity, in the dark), after which visual examination is used to quantify any observed protection.

utilizes an agar cube containing the necessary vitamins and nutrients for the healthy growth of insects without any secondary metabolites to compromise test results (Fig. 2).[14] Test substances are added to this diet at the time of formulation by coating them on metabolically inert α-cellulose·to ensure an even distribution in the medium and to prevent pooling due to incompatible solubilities. The insect is then raised exclusively on this diet from neonate through a number of instars. The time period of the assay allows observations of long term effects and, since several molts occur, information of the effects on the complex metabolism of molting can be acquired. As a no-choice bioassay, the artificial diet assay can be used with the choice leaf disk assay to give information on antifeedant strength. Weak antifeedants exhibit antifeedant activity only while a choice of diet is possible, and their effect fails when an alternative diet is unavailable. However, since situations in the natural environment and in future agricultural applications of antifeedants involve a choice

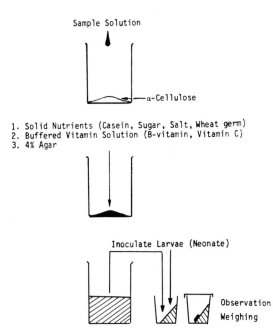

Sample Solution

—α-Cellulose

1. Solid Nutrients (Casein, Sugar, Salt, Wheat germ)
2. Buffered Vitamin Solution (B-vitamin, Vitamin C)
3. 4% Agar

Inoculate Larvae (Neonate)

Observation
Weighing

Fig. 2. The no-choice artificial diet feeding assay. Test
compounds are coated on metabolically inert α-cellulose and
dispersed into a nutrient-rich, agar diet. Portions of the
diet are placed in individual containers and neonate larvae
are raised exclusively on this diet for about two weeks
(28°, 80% relative humidity, 12-14 h day length). Effects
are measured by comparison to a control.

of diet, the use of no-choice assay systems only might over-
look interesting and valuable compounds and represent an
unreal restriction. Also, toxic antifeedants forced to be
consumed in a no-choice assay might be labeled as a toxic
component only with its antifeedant property missed. Both
assays give useful but different information.

The ether soluble fraction was further investigated
using the artificial diet feeding assay to direct the isolation. As a result the related neo-clerodane ajugarin-IV ($\underline{5}$)
was isolated; this compound possesses mild insecticidal
activity against the silkworm (Bombyx mori; 500 ppm LD_{95})
and growth inhibitory activity against the pink bollworm
(Pectinophora gossypiella; 500 ppm ED_{50}).[15] It is interest-

Table 1. Effects of phytoecdysteroids on the growth and
development of larvae of Pectinophora gossypiella.

Compound	Amount in Diet (ppm)	Effect
20-Hydroxyecdysone	35	ED_{50}[a]
	50	EI_{95}[b]
Cyasterone	25	ED_{50}
	40	EI_{95}
Ajugasterone C	14	ED_{50}
	45	EI_{95}
Ponasterone A	1	ED_{50}
	2	EI_{95}

[a] ED_{50} values are the effective doses for 50% inhibition of
growth and are the means of five determinations.

[b] EI_{95} values are the effective doses for 95% inhibition of
ecdysis resulting in death, and are the means of five
determination.

ing to note the antifeedant compounds 1, 2, 3 and 4 were
overlooked by this assay while 5 was overlooked in the
previous leaf disk bioassays. Because of the similar structure of 5 to compounds 1, 2, 3 and 4 but its different
effect, the ether soluble fraction was further purified to
look for related compounds that might help in predicting a
structure-activity relationship. Ajugarin-V (6) was isolated
as an inactive but structurally similar compound[16] and
suggests that the transdecalin portion of the adjugarins is
involved in the activities rather than the perhydrofurofuran
system as previously suggested.[17] It is also possible that
there is a synergistic effect among the groups.[18]

The ethyl acetate soluble fraction tested positive in
the artificial diet feeding assay but was negative in the
leaf disk assay. Isolation of the responsible constituents
revealed three phytoecdysteroids; 20-hydroxyecdysone (7),
cyasterone (8) and ajugasterone C (9).[19] Their effect is
quite spectacular on some of the insect species tested. At
lower levels all three compounds induced growth inhibitory
activity. At slightly higher levels in the diet, the molting
pattern was disturbed (Table 1). The molting cycle's initial

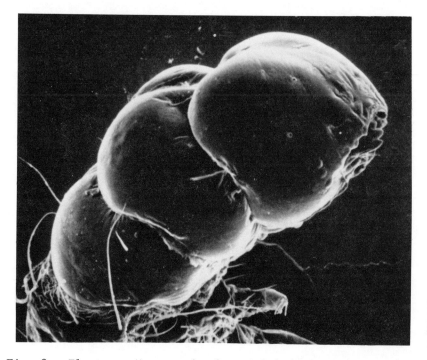

Fig. 3.   Electron micrograph of a pink bollworm, Pectinophora
gossypiella, after ingestion of a crude methanol extract of
the root bark of Ajuga remota.   This insect has three head
capsules that mask its functional mouth parts.   The insect
eventually starved to death.   Magnification X 113.

process (apolysis) appears to be unaffected (although there
can be some simulation), and the cuticular epithelium
separates from the old, overlying cuticle.   However, after
the new cuticle is formed complete, the normal process of
hydrostatic expansion of the new cuticle (ecdysis) resulting
in the tearing away of the old cuticle was radically
disrupted and caused the molting cycle to fail because of
ecdysis inhibition.   The uncompleted cycle left the old head
capsule attached to the new one and feeding was blocked
(Fig. 3).   After surgical removal of the old head capsule,
the newly formed mouth parts were seen to be deformed by
the old head capsules and this also inhibited feeding (Fig.
4).   Death occurred due to starvation although it was

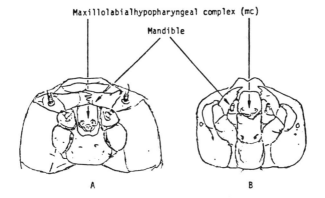

Fig. 4. Schematic depicting (ventral) dysfunctional mouth-
parts in ecdysis inhibited Bombyx mori larvae. (A) Normal
5th instar head capsule with fully closed mandibles.
(B) Ecdysis inhibited 5th instar head capsule observed after
removal of the 4th instar head capsule. The 4th instar head
capsule prevented full expansion of the 5th instar head
capsule; this resulted in the forward position of the
maxillolabial-hypopharyngeal complex (mc) and incomplete
closure of the mandibles.

possible for subsequent unsuccessful molts to create up to
three head capsules.[20] Of the five species tested B. mori,
P. gossypiella and Spodoptera frugiperda showed comparable
behavior. Heliothis zea and H. virescens were unaffected
by the phytoecdysteroids. It is not know if this is
because of detoxification or by the efficient excretion of
the phytoecdysteroids from the hemolymph. We hope to answer
this question through metabolic studies currently underway.

    Although 7, 8 and 9 showed no antifeedant activity to
lepidoptera larvae in the leaf disk assay, it was decided
that they would be suitable for testing in an antifeedant
test for aphids since they were water soluble. As aphids
feed by sucking rather than chewing, a different bioassay
method is necessary. A nutrient solution is placed in a
vial capped with parafilm to resemble the waxy cuticle of
a leaf.[21] This is then inverted and exposed to a group of
aphids (Schizaphis graminum; Biotype C green bug) (Fig. 5).
Water soluble test compounds can be dissolved directly in
the aqueous nutrient solution and the number of aphids

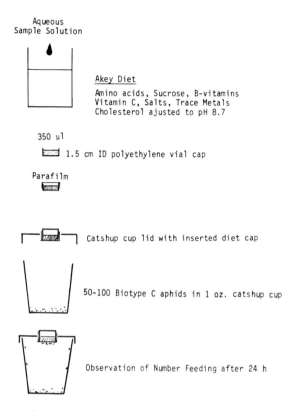

Fig. 5. Antifeedant bioassay for the greenbug, Schizaphis graminum.

feeding after 24 h are determined and compared with a control. As can be seen from Table 2, ajugasterone C is ten times more effective as an antifeedant than 20-hydroxyecdyson and over thirty times more effective than cyasterone. Howeve the three compounds are comparable in toxicity and growth inhibition (Table 1).

The original isolation of phytoecdysteroids from A. remota was by reverse phase HPLC. In order to purify large amounts of compounds for further metabolic studies, Droplet countercurrent chromatography (DCCC) was employed. It was fortuitous that during the DCCC isolation, two other compoun were also easily purified in large amounts (Fig. 6). These

Table 2. Feeding deterrency of three phytoecdysteroids on the greenbug, Schizaphis graminum.[a]

Compound	ED_{50} (ppm in diet)[b]
20-Hydroxyecdysone	650
Cyasterone	2000
Ajugasterone C	62

[a] Biotype C of S. graminum from a mixed population in a 24 h no-choice bioassay.

[b] ED_{50} is the effective dose for 50% feeding when compared to the control.

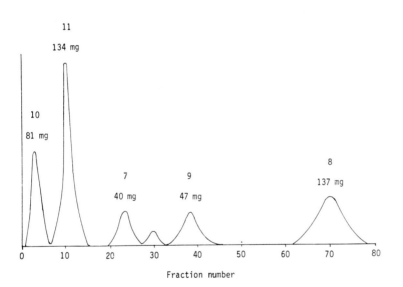

Fig. 6. DCCC separation of the ethyl acetate extract of Ajuga remota (25 g) with $CHCl_3$-CH_3OH-H_2O (13:7:4, v/v) by the ascending method. The seco-iridoid glycosides harpagide (10) and acetyl harpagide (11) were fortuitously isolated along with the phytoecdysteroids 7, 8 and 9.

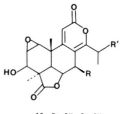

10, R= OH, R'= H
11, R= OAc, R'= H
12, R= OAc, R'= OH

13, R= OH, R= CH₃
14, R= R'= H

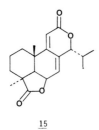

15

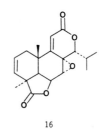

16

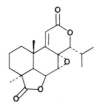

17

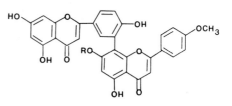

18, R= H
19, R= CH₃

bitter components were identified by spectral means as the
known iridoid glycosides harpagide (10) and acetyl harpagide
(11).[22] TLC showed a high concentration of these compounds
in the aqueous soluble fraction, and after cross-flow micro-
filtration with a stirred cell (Spectra/Por 43 mm stirred
cell, type C 1000 MWCO membrane, Spectra Medical Industries
Inc., New York, New York) was performed, almost all color
was removed. TLC of the filtered solution showed only two
spots corresponding to 10 and 11. This technique is relatively uncommon in dealing with organic type compounds but
was very useful in this instance and will probably see more
use as more labs become equipped to deal with water solubles.
Compounds 10 and 11 showed no activity in the artificial diet
feeding assay with P. gossypiella at concentrations up to
2000 ppm, nor in lettuce seed germination and growth tests.
They did show activity against aphids utilizing the above
aphid feeding assay and 10 caused an odd type of escape
response in preliminary tests on the snail Biomphalaria
glabratus. The latter is a host for the parasite which
causes schistosomiasis. When snails were placed in a solution containing 10 they would move up the beaker and leave
the water even though they are normally aquatic. Compounds
similar to 10 and 11 are known to possess a range of pharma-
ceutical activities.[23] Jaranidoside (12), which contains an
additional hydroxyl group, stimulates cardiac and smooth
muscle tissue.[24]

In summary, we have shown that A. remota contains three
distinct classes of compounds capable of affecting the
feeding behavior of lepidopteran larvae and aphids, causing
the disruption of the normal molting process of several
species of insects, causing growth inhibition and toxic
effects on insects and affecting the behavior of molluscs.

FAMILY PODOCARPACEAE

Species of the genus Podocarpus are found in Africa,
Australia, China, Chile and Japan and are considered relatively immune to insect attack. Their defense has been shown
primarily to be chemical in nature. Over forty examples of
biologically active nor- and bisnorditerpene dilactones have
been isolated from seventeen different species.[25] Several
species also contain phytoecdysteroids. Dried leaves of
P. gracillior grown in Nairobi, Kenya, were extracted in
methanol.[26] Assays of the crude methanol extract showed

molting inhibition, feeding inhibition and toxicity to
several lepidopteran larvae. As with A. remota, several
biological assays were used to direct the isolation of
chemical factors responsible for the activity. The crude
extract was initially partitioned between ethyl acetate and
water. The ethyl acetate portion was then washed with ethyl
ether to remove pigments and fractionated by column chromatography. Further fractionation and bioassays identified
nagilactone C (13) and nagilactone D (14) as being responsible for the antifeedant activity against H. zea and S.
frugiperda in the choice host plant leaf disk assay. These
compounds also proved inhibitory to growth and were lethal
in artificial diet feeding assays (Table 3). Fractionation
of the ether soluble portion gave nagilactone F (15) and the
related podolide (16) and dihydropodolide (17). Podolide
showed toxicity to P. gossypiella but both 15 and 16 showed
growth inhibition similar to the toxic nagilactones 13 and
14. The relative strength of these compounds depended on
the insect species tested (Table 4). The mode of toxicity
and growth inhibition was not tested for but may be due at
least in part to the feeding deterrent effect of these
compounds.

Table 3. Choice host plant leaf disk assay (24 hours) with
third instar larvae involving two norditerpene dilactones.[a]

Species	Test Compound	PC_{50} (μg/disk)[b]
Heliothis zea	Nagilactone C	38
	Nagilactone D	15
Spodoptera frugiperda	Nagilactone C	38
	Nagilactone D	20

[a] Podolide and nagilactone F were not tested in the leaf
disk assay; podocarpusflavone A and 7'',4'''-
dimethylamentoflavone were found to be inactive in this
bioassay.

[b] PC_{50} values are concentrations of test compounds at which
less than 5% of the treated leaves are eaten, while
greater than 50% of the untreated leaves are eaten. The
experiment was terminated when more than 50% but less
than 80% of the control leaf disks were consumed.

Table 4. Survival and growth rate activities of four
norditerpene dilactones in a 12-day no-choice artificial
diet feeding assay.

Test Compound		LD_{90} (ppm)[a]	ED_{50} (ppm)[b]
Heliothis zea	Nagilactone C	1500	20
	Nagilactone D	800	4
	Nagilactone F		30
	Podolide		12
Spodoptera frugiperda	Nagilactone C	2000	18
	Nagilactone D	2000	6
	Nagilactone F		12
	Podolide	2000	7
Pectinophora gossypiella	Nagilactone C	1500	14
	Nagilactone D	200	4
	Nagilactone F		20
	Podolide	300	9

[a] LD_{90} values are the lethal doses for 90% death and are
the means of five determinations.

[b] ED_{50} values are the effective doses for 50% growth inhibition and are the means of five determinations.

Later fractions (eluted after 13 and 14) of the ethyl
acetate portion of the crude extract contained non-toxic
growth inhibitory activity and molting inhibition activity.
Two biflavonoids, podocarpusflavone A (18) and 7'',4'''-
dimethylamentoflavone (19) were purified and shown to possess
non-toxic growth inhibition activity (Table 5). This was
the first observation of inhibition of insect growth by
biflavonoids. Their mode of action is unknown although the
leaf disk assay suggests that the growth inhibition is not
due to an antifeedant effect. On a molal basis, their
activities as growth inhibitors against H. zea[27] are comparable to monoflavonoids.

The fractions exhibiting toxic ecdysis inhibitory
activity were combined to yield crystal needles of the
phytoecdysteroid Ponasterone A (20). Compound 20 showed

Table 5.   Growth inhibitory activity of two biflavonoids in the no-choice artificial diet feeding assay.[a]

Species	Test Compound	ED_{50} (ppm)[b]
Heliothis zea	Podocarpusflavone A	625
	7'',4'''-dimethylamentoflavone	2500
Spodoptera frugiperda	Podocarpusflavone A	1400
	7'',4'''-dimethylamentoflavone	3300
Pectiophora gossypiella	Podocarpusflavone A	2300
	7'',4'''-dimethylamentoflavone	2300

[a] No deaths were recorded for either biflavonoid to 5000 ppm.

[b] ED_{50} values are the doses required for 50% inhibition of growth and are the means of five determinations.

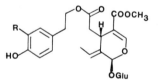

21, R= OH
22, R= H

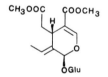

23

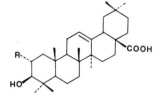

24, R= H
25, R= OH

molting hormone activity when fed to P. gossypiella and when
orally injected into B. mori. When compared to ecdysteroids
7, 8 and 9 from A. remota, 20 had over a ten-fold greater
effect on the growth and development of P. gossypiella in
an artificial diet assay (Table 1). Although 7 is an actual
insect molting hormone, it is believed 20 is more effective
in disrupting molting because of its stronger binding to
proteins.[28]

P. gracillior contains three distinct classes of
compounds capable of a multidirectional chemical defense.
Nagilactones 13 and 14 and compound 16 represent antifeedant.
The biflavonoids 18 and 19 act as growth inhibitors with
neither antifeedant activity nor toxic effects. The poly-
hydroxysteroid 20 has strong growth inhibitory and ecdysis
inhibitory activity. It is also apparent that the
nagilactones play a role in the germination of potential
plant competitors.[25]

FAMILY OLEACEAE

The bitter olive Olea europaea has been used for centu-
ries as a source of food and oil. It has been known to be
resistant to microbe attack for some time. Our preliminary
screening of the methanol extract of fruits from trees grown
in Berkeley, California showed antimicrobial activity
against the gram positive Bacillus subtilis. The crude
methanol extract was partitioned between ethyl ether, ethyl
acetate and water in a manner similar to that employed in the
study on A. remota.[29] Microbial assays showed the active
components to be in the ether soluble portion. Repeated
attempts to purify the active compounds failed and it
appeared that the active compounds were too unstable to
purify. Since the crude extract contained a few major
compounds, it was decided to look more closely at these and
abandon for a time this bioassay directed portion of the
extracts. Since the main compounds were polar DCCC was used
to purify the components. Final purification was accomplished
after two sequential DCCC runs (Fig. 7). The bitter seco-
iridoid glycosides oleuropein (21) and ligstroside (22) were
isolated in a pure state and in large amounts. The literature
disclosed that oleuropein and its enzymatic hydrolysis
products exhibited antimicrobial activity against bacteria
and yeast.[30] We predicted that the hydrolysis products of
21 and 22 would be ether soluble and unstable. Enzymatic

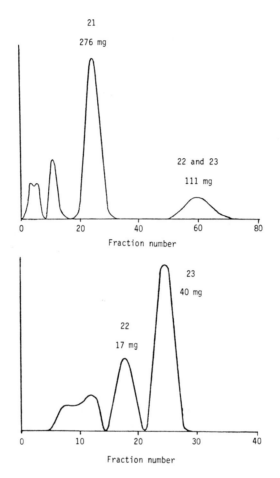

Fig. 7. Top: DCCC of the methanol extract of Olea europaea (1.0 g) by the ascending method with CHCl₃–CH₃OH–H₂O (13:7:4, v/v); Bottom: DCCC of the mixture of ligstroside (22) and its dimethyl artifact (23), (87 mg from the previous run) by the ascending method with C_6H_6–CHCl₃–CH₃OH–H₂O (5:5:7:2, v/v).

hydrolysis in vitro with β-glucosidase confirmed the insta-bility of the aglycone. Since the plant is known to contain a β-glucosidase which would hydrolyze the glucose moiety, it was believed that the aglycone would be the active component which could not be purified. When 21 and 22 were tested

Table 6. Antimicrobial activities by the paper disk method in the presence of β-glucosidase.[a]

	mg/disk	Oleuropein (21)			Ligstroside (22)	
		2.0	1.0	0.5	1.0	0.5
Bacillus	pH 8	trace	-[b]	-	8	7
subtillis	pH 7	14	11	8	16	12
	pH 5	NG[c]	NG	NG	NG	NG
Saccharomyces	pH 8	25	20	14	-	-
cerevisiae	pH 7	16	13	-	-	-
	pH 5	-	-	-	-	-
Escherichia	pH 8	-	-	-	-	-
coli	pH 7	-	-	-	-	-
	pH 5	-	-	-	-	-

[a] No zones of inhibition were observed in any cases without added β-glucosidase.

[b] Indicates no zone of inhibition.

[c] Indicates that the microorganism did not grow in the medium.

against Saccharomyces cerevisiae (yeast), B. subtilis (gram positive) and Escherichia coli (gram negative) by the paper disk method, neither oleuropein nor ligstroside showed activity. However, when the assays were repeated in the presence of β-glucosidase, both compounds gave a zone of inhibited growth of B. subtilis around the paper disk at pH 7. No activity was observed with either compound on E. coli, but 21 proved active against the yeast S. cerevisiae at pH 7 and pH 8 (Table 6). It seems reasonable that the active aglycone could be present in small amounts and replaced when needed by hydrolysis of oleuropein. It is also possible that the active compound of the ether soluble material is actually the aglycone of oleuropein and/or ligstroside; however, this has yet to be established. When 21 and 22 were tested against Biomphalaria glabratus, the South American aquatic snail responsible for spreading schistosomiasis, both were found active (21, LD_{50} = 250 ppm; 22, LD_{50} = 100 ppm).[31]

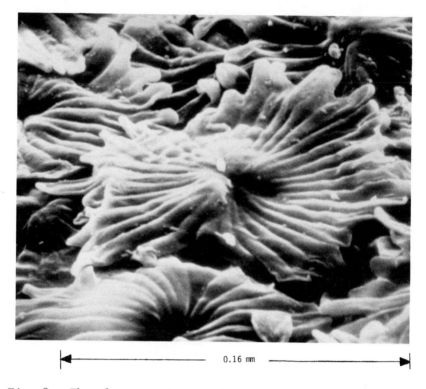

|←——————————————— 0.16 mm ————————————————→|

Fig. 8. The electron micrograph of the abaxial surface of
a leaf of Olea europaea. Magnification X 550.

        Observation of the leaf showed a deposit of crystalline
material which covered the upper and lower leaf surface. An
electron micrograph of the abaxial surface is shown in
Figure 8. Since the first barrier to fungal or other microbe
attack is the leaf cuticle, the components of this coating
were investigated. TLC showed the coating to be the triter-
pene oleanolic acid (24) containing a slight amount of
maslinic acid (25). These had been found in extracts of
whole leaves except it was not previously realized that they
came from the leaf surface. Oleanolic acid did not exhibit
antimicrobial activity. However, as an effective hydrophobic
barrier on the leaf surface, it does keep water from
collecting. The latter is important for the germination
of microorganisms and the thick hydrophobic barrier may
slow the penetration of successfully germinated microbes

through the cuticle surface to the enclosed nutrients.[33] The thickness of this barrier is large enough to keep certain sucking insects such as aphids from penetrating into the leaf.

The olive exhibits forms of chemical defense of a more cryptic nature than discussed in the previous two examples. It is possible that oleuropein and ligstroside are phyto-alexin precursors. The production of phytoalexins must be turned on in some manner before their antibiotic activities can be observed. Both compounds are also active against at least one species of mollusk. Traditional testing of oleanolic acid or maslinic acid would have dismissed them as uninteresting while as a thick coating, their chemical properties would be perceived as a physical barrier stable to the environment. Other plants utilize waxy surfaces in a similar manner. However, most waxes are composed of mixtures of long chain hydrocarbons.[34]

CONCLUSIONS

We have discussed examples from three families of plants. One point we have tried to emphasize is the diver-sity of plant chemistry and the varied defenses these compounds represent not just amongst the plant kingdom but within a single species. Although the examples were not exhaustively investigated, the use of relatively simple and inexpensive bioassays made possible the isolation of a wide range of compounds. DCCC was of particular use as it enabled the isolation of highly polar compounds that might have been discarded if more conventional solid support chromatographies has been employed exclusively.

As we explore we find that many plants are not as passive in self-defense as they seem. The evolved defen-sive strategy of plant species provides a variety of unique effects on the harsh environment in which plants exist. The discovery of their tactics continues to give a better picture of the complex ecological relationships between plants and their surroundings. The time period that these relationships have survived attest to their superb balance within the ecological system. As we continue to encounter problems with the more traditional means of pest control, it will be to our benefit to model our future efforts on these balanced tactics already established through evolutionary pressures.

REFERENCES

1. WILLIAMS, K., L. GILBERT. 1981. Insects as selective
   agents on plant vegetative morphology: Egg mimicry
   reduces egg laying by butterflies. Science 212:
   467-469.
2. JUNIPER, B.E., C.E. JEFFREE. 1983. Plant Surfaces.
   Edward Arnold Publishers Limited, London, 42 pp.
3. BERNAYS, E.A., R.F. CHAPMAN. 1978. Plant chemistry
   and acridoid feeding behaviour. In Biochemical
   Aspects of Plant and Animal Coevolution. (J.B.
   Harborne, ed.), Academic Press, New York, pp. 99-137.
4. KUBO, I., K. NAKANISHI. 1977. Insect antifeedants and
   repellents from African plants. In Host Plant
   Resistance to Pests. (P.A. Hedin, ed.), ACS Symposium Series 62, American Chemical Society, Washington,
   D.C., pp. 165-178.
5. HARBORNE, J.B. 1982. Introduction to Ecological
   Biochemistry. Academic Press, New York, pp. 66-67.
6. KUBO, I., J.A. KLOCKE. 1981. Azadirachtin, insect
   ecdysis inhibitor. Agric. Biol. Chem. 46(7): 1951-
   1953.
7. KUBO, I., J.A. KLOCKE. 1981. Limonoids as insect
   control agents. Les Colloques de I'INRA 7: 117-129.
8. NAKANISHI, K. 1975. Structure of the insect antifeedant azadirachtin. In Recent Advances in
   Phytochemistry. (V.C. Runeckles, ed.), Vol. 9,
   Plenum Press, New York, pp. 283-298.
9. CREMLYN, R. 1979. Pesticides. John Wiley and Sons,
   New York, pp. 41-42.
10. FEENY, P. 1976. Plant apparency and chemical defense.
    In Recent Advances in Phytochemistry. (J.W. Wallace,
    R.L. Mansell, eds.), Vol. 10, Plenum Press, New York,
    pp. 1-40.
11. KUBO, I., Y. LEE, V. BALOGH-NAIR, K. NAKANISHI, A.
    CHAPYA. 1976. Structure of ajugarins. J. Chem.
    Soc., Chem. Commun., pp. 949-950.
12. KUBO, I., M. KIDO, Y. FUKUYAMA. 1980. X-ray crystal
    structure of 12-bromoajugarin I and conclusion on
    the absolute configuration of ajugarins. J. Chem.
    Soc., Chem. Commun., pp. 897-898.
13. KRIEGEN, J.H. 1982. Chemistry confronts global food
    crisis. Chem. Eng. News, December 20, pp. 9-23.
14. CHAN, B.G., A.C. WAISS, JR., W.L. STANLEY, A.E.
    GOODBAN. 1978. A rapid diet preparation method for
    antibiotic phytochemical bioassay. J. Econ. Entomol.
    71: 366-368.

15. KUBO, I., J.A. KLOCKE, T. MIURA, Y. FUKUYAMA. 1982. Structure of ajugarin IV. J. Chem. Soc., Chem. Commun., pp. 618-619.
16. KUBO, I., Y. FUKUYAMA, A. CHAPYA. 1983. Structure of ajugarin V. Chem. Lett., pp. 223-234.
17. KOJIMA, Y., N. KATO. 1979. Synthesis and conformational analysis of the perhydrofuro [2,3-b] furan compounds by using lanthanide shift reagent and empirical force-field calculations. Tetrahedron Lett. 48: 4667-4670.
18. LUTEIJN, J.M., A. de GROOT. 1981. 9α-(acetoxymethyl)-8α, 8'-epoxy-3α,4,4-trimethyl-trans-decalin-1α-ol acetate, a model for the investigation of structure-activity relationships of the insect antifeedant neoclerodanes. J. Org. Chem. 46: 3448-3452.
19. KUBO, I., J.A. KLOCKE, I. GANJIAN, N. ICHIKAWA, T. MATSUMOTO. 1983. Efficient isolation of phytoecdysones from Ajuga plants by high-performance liquid chromatography and droplet counter-current chromatography. J. Chromatog. 257: 157-161.
20. KUBO, I., J.A. KLOCKE. 1983. Isolation of phytoecdysones as insect ecdysis inhibitors and feeding deterrents. In Plant Resistance to Insects. (P.A. Hedin, ed.), ACS Symposium Series 208, American Chemical Society, Washington, D.C., pp. 329-346.
21. DREYER, D.L., J.C. REESE, K.C. JONES. 1978. Aphid feeding deterrents in sorghum: bioassay, isolation, and characterization. J. Chem. Ecol. 7: 273-284.
22. KUBO, I. unpublished results.
23. STITCHER, O. 1977. The analysis of iridoid drugs. Pharm. Acta. Helv. 52(1-2): 20-32.
24. CHUNG, B., H. LEE, J. KIM. 1980. Iridoid glycoside. I. Studies on the iridoid glycoside of Ajuga spectabolis (Nakai). Saengyack Hak hoe Chi. 11(1): 15-23.
25. ITO, S., M. KODAMA. 1976. Norditerpene dilactones from podocarpus species. Heterocycles 4(3): 595-624.
26. KUBO, I., T. MATSUMOTO, J.A. KLOCKE. 1984. Multichemical resistance of the conifer (Podocarpus gracilior (Podocarpaceae) to insect attack. J. Chem. Ecol. 10(4): 547-559.
27. KUBO, I., J.A. KLOCKE, T. MATSUMOTO. 1983. Identification of two insect growth inhibitory biflavonoids in Podocarpus gracilior. Rev. Latinoamer. Quim. 14(2): 59-61.
28. KOOLMAN, J. 1982. Ecdysone metabolism. Insect Biochem. 12(3): 225-50.

29. KUBO, I., A. MATSUMOTO, I. TAKASE. 1984. A multi-
     chemical defense mechanism of bitter olive Olea
     europaea (Oleaceae): Is oleuropein a phytoalexin
     precursor. J. Chem. Ecol. 11(2): 251-263.
30. ASAKA, Y., T. KAMIKAWA, T. KUBOTA, H. SAKAMOTO. 1972.
     Structures of seco-iridoids from Ligstrum
     obtasifolium (Steb. et Zucc.). Chem. Lett. 00:
     141-144.
31. KUBO, I., A. MATSUMOTO. 1984. Molluscicides from
     olive Olea europaea and their efficient isolation
     by countercurrent chromatographies. J. Agric. Food
     Chem. 32: 687-688.
32. RONCERO, A.V., M.L. JANER. 1969. Acidos triterpenicos
     del olivo. Grasas Aceites 20: 133-138.
33. HARBORNE, J.B., J.L. INGHAM. 1978. Biochemical
     aspects of the coevolution of higher plants with
     their fungal parasites. In J.B. Harborne, ed.,
     op. cit. Reference 3, pp. 343-400.
34. JUNIPER, B.E., C.E. JEFFREE. 1983. Plant Surfaces.
     Edward Arnold Publishers Limited, London, 6 pp.

Chapter Eight

PHEROMONAL COMMUNICATION BETWEEN PLANTS

DAVID F. RHOADES

Department of Zoology
University of Washington
Seattle, Washington 98195

INTRODUCTION

It has become generally accepted that plant secondary metabolites constitute an important defense of plants against attack by herbivores and other consumers.[1-3] It is also becoming clear that plants are far from static resources to herbivores, as had been widely assumed in the past. Plant nutritional quality to herbivores can vary with the degree of both physical and biotic stress experienced by plants. Physical stresses such as drought, waterlogging, depletion of soil nutrients, frost damage, and pollution often increase susceptibility of plants to herbivores.[4-6] In physically stressed plants, this increased susceptibility appears to be due to a combination of increased content of amino acids and other nutrients,[7,9] and lowered commitment to defense.[5,6] Reports of decreased nutritional quality to herbivores, as measured in bioassays

195

or increased content of defensive compounds, in plants
previously attacked by herbivores or experiencing current
attack, are becoming increasingly common.[6,10,11,17]
Defensive responses of plants to attack by herbivores have
been reported in as many as 29 species in 13 families
including herbs, grasses, a sedge, a vine and trees, so
these responses may be universal.

Changes in plant chemistry caused by physical and
biotic stresses are potentially very important to herbivore
population dynamics. Population models based on competition, predation, parasitism, disease, and weather effects
acting directly on herbivores have not proven effective in
predicting future herbivore population trends in most
cases.[12-15] It is probable that successful explanatory
and predictive herbivore population models will be obtained
only when the effects of biotic and physical stress on
plant nutritional quality and chemical defenses are also
taken into account.[6,10,16,17] Information transfer from
attacked to unattacked plants is possibly also important
in herbivore population dynamics and plant-herbivore
interactions in general.

Since 1979, field experiments have been conducted which
are designed to detect changes in the nutritional quality
or red alder (<u>Alnus</u> <u>rubra</u> Bong, Betulaceae) and Sitka
willow (<u>Salix</u> <u>sitchensis</u> Sanson, Salicaceae) trees, induced
by insect attack. In these experiments, colonies of western
tent caterpillars (<u>Malacosoma</u> <u>californicum</u> <u>pluviale</u> Dyar,
Lasiocampidae) or fall webworms (<u>Hyphantria</u> <u>cunea</u> Drury,
Arctiidae) were placed on trees while periodically moni-
toring leaf quality by insect feeding bioassays and chemical
analysis. Larvae fed leaves from alder trees experiencing
attack by tent caterpillars showed lowered growth rates and
other fitness parameters compared to those fed leaves from
unattacked control alders. The attacked trees also showed
increased proanthocyanidin content of leaves when compared
to the control trees.[18] These changes in leaf quality of
the attacked trees occurred within 27 days of the initiation
of attack. Similarly, attack of 'naive' (see Experiment V
below) Sitka willows by webworms resulted in lowered growth
rates of assay webworms.[18] However, many of the experiments
with Sitka willows gave results suggesting that both the
test and unattacked control trees were changing their leaf
quality in response to placement of insects on the test
trees. This implied that unattacked plants receive and

respond to signals from nearby plants experiencing attack
by insects and during these experiments an overall picture
strongly suggestive of communication among plants gradually
emerged.

The experiments which led to this conclusion will be
described in some detail.

MATERIALS, METHODS, AND EXPERIMENTAL RESULTS

In all the following experiments the boles of test and
control trees were treated with tanglefoot gum spread on
aluminum foil wrapped around the boles to prevent migration
of larvae from the test trees to the control trees. In
addition, branches (10-30% total foliage volume) on each
test and control tree were similarly treated with tanglefoot
or enclosed in netting bags to exclude caterpillars. Leaves
were then periodically removed from these protected branches
during the experiment for bioassay and chemical analysis in
the laboratory. In this way, it was ensured that any changes
in quality of leaves of attacked trees resulted from a
systemic response in the plant rather than preferential
consumption of high quality leaves by the caterpillars.

Leaves were removed from assay branches by breaking
them off at their petioles, transported to the laboratory in
polyethylene bags at room temperature, and fed ad lib. to
groups of larvae (4-20 insects, one replicate per tree) which
were maintained in petri dishes (16 hours light, 25°C; 8
hours dark, 20°C). The number of larvae per replicate, and
their instar, varied between assays depending on availability,
but for any given assay equal numbers of larvae of similar
size and instar were randomly assigned to test and control
replicates. Fresh leaves were homogenized with methanol/
water (85:15, v/v) at room temperature within 5 hours of
collection. The extracts were filtered, stored in amber
glass bottles at 5°C until analysis. Proanthocyanidins in
the extracts were determined within 3 days by the method of
Swain and Hillis.[19]

In Experiment I, leaf damage was determined by counting
the total number of leaf clusters on the trees and during
the experiment periodically counting the number of attacked
clusters. In the other experiments leaf damage was estimated
following each experiment by randomly selecting 3 leaf

clusters at each of 12 stations on each test and control
tree. The stations were distributed throughout the crown
with respect to both compass direction and height within
the crown (north high, north intermediate, north low, etc.).
For each of the clusters at a given station the percent leaf
area missing for each leaf (ca. 10-35 leaves per cluster)
was visually estimated to the nearest 5%. Similarly, on
each assay branch damage was estimated for all leaves on
each of four randomly selected clusters distributed north,
west, south, and east. Average percent leaf damage for each
tree was then calculated using a visual estimate of the
fraction of the total foliage volume constituted by the
assay branches. Leaf area lost to sampling was calculated
from the number of leaves sampled, the average number of
leaves per cluster, the number of clusters per assay branch,
and the visual estimate of assay foliage fraction.

Experiment I - 1979

From a group of 20 Sitka willow trees (average height
3.6 ± 0.1 [S.E.] m and volume 1.2 ± 1.0 m³), 10 test and 10
control trees were randomly assigned in pairs. The distance
between each control tree and the nearest test tree averaged
3.5 ± 0.7 (S.E.) m. Two tent caterpillar egg masses, containing an average of 193 viable eggs per mass, were placed
on each of the test trees on March 20 and these hatched during
early April. Proanthocyanidin (PA) contents of leaves from
all trees and leaf damage by the tent caterpillars were
periodically monitored (Fig. 1). Although little difference
in leaf PA content of both test and control trees was
observed throughout the season, a maximum in PA content of
both test and control trees coincided with maximum damage to
the test trees by the caterpillars (Fig. 1). Following
disappearance of the caterpillars from the trees, due to
pupation and mortality, leaf PA content of both test and
control trees dropped to a lower value. Test and control
trees received little damage from other insects during
the experiment.

Experiment II - 1979[18]

Tent caterpillar egg masses were placed on 7 Sitka
willow trees but not on 7 controls. The distance between each
control tree and the nearest test tree averages 3.5 ± 0.4
(S.E.) m. Starting 17 days after the field insects hatched,
additional tent caterpillars were raised in the laboratory

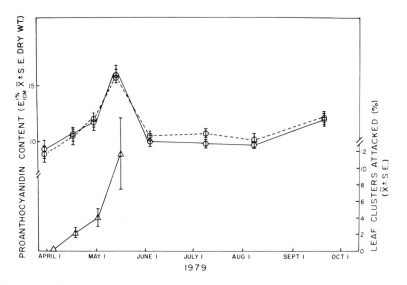

Fig. 1. Top: Leaf proanthocyanidin contents of 10 Sitka
willow trees attacked by western tent caterpillars (——)
and 10 unattacked control willows (----) on various dates.
Bottom: Leaf clusters (%) on the attacked trees that had
been damaged by tent caterpillars on various dates.

on periodically replaced leaves detached from assay branches
of attacked and control trees (Fig. 2). No differences in
growth between test and control assay insects were observed,
so the biomass of attacking insects was greatly increased by
placing additional colonies of tent caterpillars on the test
trees (reload, Fig. 2). Following the reload operation, a
large drop in biomass, due to increased mortality and cur-
tailed growth, was observed for both test and control assay
insects (Fig. 2).

The results of these two experiments in 1979 suggested
the possibility that unattacked willows could detect and
respond to signals from nearby trees attacked by tent caterpillars. Excavation of 20 Sitka willows of the same age as
the study trees, at the same site, gave no evidence of root
connections and as trees in general[20] and willows in particular[21] emit volatile terpenes and other compounds of unknown
function into the atmosphere, it was concluded that if communication occurred, it was possibly via airborne substances.

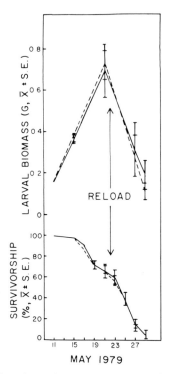

Fig. 2. Biomass (top) and survivorship (bottom) of western
tent caterpillars (25 larvae per replicate, one replicate
per tree) fed leaves from 7 Sitka willow trees attacked by
tent caterpillars (——) and 7 unattacked control willows
(----). Additional tent caterpillars were placed on the
trees (reload) on the indicated date (see text).

Experiment III - 1980

        In spring of 1980, 32 Sitka willow trees of average
height 5.6 ± 0.2 (S.E.) m and volume 4.1 ± 0.6 m^3 were
selected, consisting of a test group of 8 trees (T), two
nearby control groups of 8 trees each (C1 and C2) at a
distance of about 3-10 m from the test group, and a distant
control group of 8 trees (C3) at a distance of about 100 m,
judged to be possibly out of range of any pheromonal effects.
A trench 1.1 m in depth was excavated between groups T and
C1 to minimize possible root contact between these plants.
On May 21, colonies of tent caterpillars were placed on the

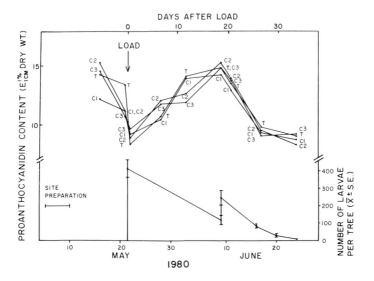

Fig. 3. Top: Average leaf proanthocyanidin contents of four groups of willow trees (n = 8 in each group) consisting of a test group (T), loaded with western tent caterpillar colonies on the indicated date, a nearby control group (C1) separated from group T by a trench, another nearby control group (C2) and a more distant control group (C3). Bottom: Numbers of caterpillars attacking the group T trees on various dates. Colonies were placed on the group T trees on May 21 (LOAD) and additional colonies were placed on the group T trees on June 9 (see text).

test trees (load) and the number of caterpillars remaining on the trees was censused periodically thereafter (Fig. 3). Additional caterpillars were added on June 9.

No significant difference in growth rates or survival were observed between groups of laboratory tent caterpillars fed detached leaves from the four groups of trees throughout the experiment. Leaf PA content of all groups was high and declining before load. Following load all groups of trees slowly increased in PA content to a maximum, then decreased to a low value coincident with decreasing density of attacking insects (Fig. 3).

These results, together with those of Experiment I were consistent with an induced increase in leaf PA content of all groups of trees caused by placing caterpillars on the test trees. The results also implied that the signals were probably airborne with a range of at least 100 m. The reason that leaf PA content was high and falling at the start of the experiment is unknown (Fig. 3). It could have been caused by site preparation which involved felling all trees touching the study trees (within ca. 2 m) to prevent emigration of caterpillars from test trees.

It was realized at this time that if communication occurred at distances of 100 m or greater, the existence of the phenomenon would be difficult to prove conclusively in the field with available facilities. Ideally such a study would use many different sites widely separated from each other. Half of these sites would be randomly assigned as test and the other as control. Leaf quality of attacked and their nearby control trees at each test site would then be compared to those of trees at the control sites. In this way differences or changes in leaf quality caused by any extraneous site-specific effects during the experiment could be statistically eliminated. However, such study sites were not available. Therefore several experiments were undertaken in which leaf quality of attacked and nearby unattacked trees at a single site were compared to that of trees at a distant control site. It was reasoned that while these experiments would not prove communication, they would at least provide evidence consistent or inconsistent with communication.

Experiment IV - 1981[18]

A comparison was made of leaf quality of 10 willows attacked by tent caterpillars and 10 nearby controls (attacked and control willows randomly assigned in pairs) to that of 20 willows situated at a distance of 1.5 km. In this experiment a different bioassay of leaf quality was used than that described previously. Instead of measuring the growth of larvae fed periodically replaced leaves from each group of trees over the entire course of the experiment, a series of short-term "spot" assays were performed. Weighed larvae were raised for a period of about 22 hours, after which time they were reweighed and discarded. Fresh larvae were then used in subsequent assays. This method greatly facilitated the detection of rapid changes in leaf quality.

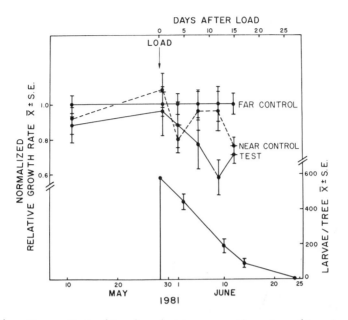

Fig. 4. Top: Normalized relative growth rates (fresh weight) of western tent caterpillars fed leaves (one repli- cate per tree) from 10 test Sitka willow trees loaded with western tent caterpillar colonies on the indicated date, 10 near control trees, and 20 far control trees. Bottom: Number of tent caterpillar larvae attacking the trees on various dates (see text).

No differences in leaf quality (as measured by fresh weight growth of the assay insects) were found among the three groups of trees prior to load or in three assays subsequent to load, but in the 5th assay, 11.5 days after load, the assay insects grew more slowly on leaves from the test trees than on leaves from the near control and distant control groups (Fig. 4). Three days later, the assay larvae grew more slowly on leaves from both the test and near control groups than on leaves from the far control group (Fig. 4).

This result is the first concrete evidence that attack of Sitka willow by tent caterpillars can induce changes in leaf quality of both attacked and nearby unattacked trees. It is difficult to imagine an extraneous site-specific

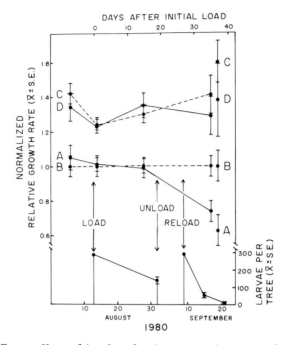

Fig. 5. Top: Normalized relative growth rates (fresh
weight) of fall webworm larvae fed leaves (one replicate
per tree) from four groups of Sitka willow trees A-D (10
trees per group) on various dates in 1980. Colonies of
webworm larvae were placed on the group A trees (LOAD),
removed from them (UNLOAD), and fresh colonies were replaced
on them (RELOAD) on the indicated dates. Connected points
represent laboratory growth rates. Unconnected points at
the right represent growth rates of larvae placed in netting
bags on the trees. Bottom: Number of larvae attacking the
trees at various dates (see text).

effect which could cause altered leaf quality, first in the
attacked trees, then in both attacked and nearby unattacked
trees at the test site compared to trees at the distant
control site, given that attacked and nearby unattacked
trees were randomly assigned.

Experiment V - 1981

In 1980, altered leaf quality had been observed in
bioassays of 10 Sitka willows (Group A) attacked by fall
webworms, compared to 10 nearby control willows (Group B)
at a distance of about 60 m, and 20 distant control willows
(Groups C and D) at a distance of 8 km (Fig. 5).[18] No
evidence for change in leaf quality of the nearby (60 m)
control trees was obtained during this first year of attack
by fall webworms.

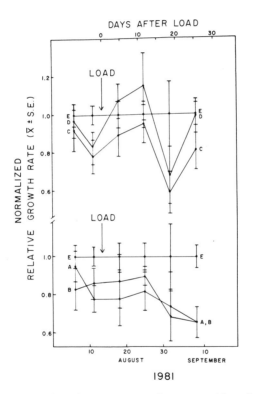

Fig. 6. Normalized relative growth rates (fresh weight) of
fall webworm larvae fed leaves (one replicate per tree)
from five groups of Sitka willow trees A-D (10 trees per
group) and E (20 trees) at various dates in 1981. Colonies
of webworm larvae were placed on groups A and C trees on
the indicated date (LOAD) (see text).

In 1981 the same trees (Group A, Fig. 6) were again
loaded with webworms. Ten of the willows at the distant
site (Group C) were also loaded with webworms, this being
their first year of attack. In addition, 20 willows
(Group E) at yet another site more than 1.5 km from both
of the other two sites were included in the experiment.
The willows in Group E were not attacked by webworms in
either year. Leaf quality of all five groups of willows
were bioassayed (short-term) with webworms prior to and
subsequent to load. For clarity the results have been
divided into two graphs (Fig. 6) in which leaf qualities of
trees at each of the two attacked sites are compared to that
of the Group E trees (not attacked in 1980 or 1981). However, statistical analyses of the growth data on each date
were performed by ANOVA which simultaneously compares the
performance of all five assay groups A-E.

There were no significant differences in leaf quality,
as measured by growth rates of assay insects, among all
groups for the two assays before load and in the first
three assays after load (Fig. 6; $p < 0.2$ for the second assay,
the most significant case). For the final assay, 25.5 days
after load, ANOVA revealed significant differences among
the five groups of trees ($p < 0.005$). A Newman-Keuls multiple
range test showed that in the final assay the leaf qualities
of Groups A (second year of attack) and B (near control
group of A) were similar and both lower than that of Group
E (unattacked in either year) and Group D (near control
group of Group C) ($p < 0.025$). Group C (first year of attack)
was not statistically separable from any of the other
groups.

Apparently, during the second year of webworm attack,
both attacked willows (Group A) and unattacked willows
(Group B), at a distance of about 60 m, can respond.

Experiment VI - 1982

In 1982 the trees of Group A were again loaded with
webworms (load 1, Fig. 7) at approximately the same density
of caterpillars per unit of tree volume as in 1980 and 1981
(ca. 55 caterpillars m^{-3}). In 1982 this amounted to about
three colonies or 540 caterpillars per tree. This was the
third consecutive year of attack for the Group A trees.
Group B (ca. 60 m from Group A) again served as near control
as it had in 1980 and 1981. These trees had not been

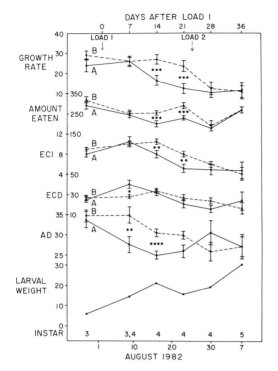

Fig. 7. Growth, consumption, and utilization parameters of fall webworm larvae fed leaves from Groups A and B trees (one replicate per tree) on various dates in 1982. Colonies of webworms were placed on the Group A trees (LOAD 1) and additional colonies were introduced into the site (LOAD 2) on the indicated dates (see text). Growth rate is expressed as % dry weight increase in mass (from the initial mass) per day; amount eaten as g dry weight per 100 g initial dry weight of larvae per day; ECI, ECD, and AD as %.[22] Larval weight (mg dry weight/larva) and instar refer to the insects used in the assays at each date. Bars indicate standard error, xxxx = p<0.001, xxx = p<0.01, xx = p<0.02, x = p<0.05 (Student's t test, two-tailed).

attacked by webworms in 1980 and 1981 and probably not during their lifetime. The intention had been to use Groups C and D, 8 km distant, as far control groups, but unanticipated road-building operations interrupted the experiment at this site.

A more elaborate bioassay of leaf quality was used in this experiment than had been used previously. Dry weight growth of webworm larvae (short-term, 18-22 hours) fed leaves from the test and control trees was measured together with amount eaten per unit of caterpillar mass per unit time, efficiency of conversion of ingested food to larval biomass (ECI), efficiency of conversion of digested food to larval biomass (ECD), and approximate leaf digestibility (AD) (Fig. 7).[22] Fresh weight growth rates (not displayed) were more variable but showed the same trends and significant differences as the dry weight growth rates displayed in Figure 7. Weights and instars of assay larvae varied from assay to assay (Fig. 7), as they have in all the short-term bioassays. Growth, feeding rate, and conversion efficiency can vary with larval weights and instars,[22] so horizontal comparisons in Figure 7 should be undertaken with caution.

In the first assay prior to load 1, there were no significant differences in any of the assay insect performance measures (Fig. 7). In the second assay, seven days after load, there was a large decrease in digestibility (AD) of leaves of the attacked trees (Group A) compared to the control trees (Group B). But, this was associated with an increase in efficiency of conversion of digested material (ECD) of leaves from the attacked trees. These two changes in leaf quality together resulted in no difference in efficiency of conversion of ingested material (ECI) between attacked and control groups. There were also no differences in amounts eaten, so no differences in growth rates were observed.

In the third assay, 14 days after load, AD of the attacked plants was again lower than that of controls. ECD of the attacked plants was the same as that of the controls (and remained so throughout the rest of the experiment), so ECI was lower for the attacked plants than for the controls. In addition, the amount eaten was lower for the attacked plants. Lowered ECI and lowered amount eaten together resulted in a lowered growth rate for the assay larvae feeding on leaves from the attacked plants than for those feeding on controls. This pattern was repeated, with minor differences in the 4th assay, 21 days after load. The results to this point were consistent with a defensive response of the attacked trees which caused their leaves to become less digestible to the insects and the insects to feed more slowly. There was no evidence suggesting changes

in leaf quality of the control trees in response to attack
of the test trees.

After the 4th assay (with the fresh weight growth data
in hand), it was decided to test the hypothesis that a
further introduction of webworms onto trees in the vicinity
of the control (B) trees would cause a reduction in their
leaf quality to that of the A trees. This hypothesis was
formulated bearing in mind the results of Experiment II
(above, Fig. 2) in which dramatic and rapid decreases in
growth and increased mortality of assay insects fed leaves
from unattacked (and attacked) willows were associated with
a sudden increase in the number of attacking insects at the
study site. So, two days after the 4th assay, an additional
81 colonies of webworms (approximately 13,000 larvae) were
placed on 10 Sitka willow trees situated between the A and B
groups, approximately 15 m from the B group (load 2, Fig. 7).
In the 5th and 6th assays, following load 2, there were no
significant differences in any of the measures of leaf
quality between Groups A and B (Fig. 7). This is consistent
with a response in the unattacked control (B) trees caused
by our introduction of additional colonies to the site closer
(15 m) to the B trees than those of the original load (60 m).

DISCUSSION

Overview of Experiments I-VI

The results of Experiments I-VI are consistent with the
idea that attack of Sitka willow by insects can lead to
changes in the leaf quality of unattacked Sitka willows up
to 100 m from the attacked trees. Although it can be argued
that in each individual experiment it was possible that the
observed changes in leaf quality of unattacked trees was
due to extraneous events (e.g. weather changes) which just
happened to coincide with the manipulation, this is unlikely
to have occurred in all six experiments. There were no
obvious changes in weather associated with changes in leaf
quality in any of the experiments, and examination of weather
records for the area reveals no correlations. Accumulated
leaf area loss, due to sampling, average 3%, 5% and 3% for
Experiments IV-VI respectively, with no significant differences among test and control trees in each experiment. In
these experiments, leaf area losses from the attacked trees
due to the insects averaged 16%, 25%, and 36% respectively.

Sampling pressure was highest (10%) compared to insect attack (30%) during the first year of attack by fall webworms, and in that year a response was seen only in the attacked trees (Fig. 5). Thus, it is unlikely that the observed changes in leaf quality of control trees were due to sampling.

All experiments using tent caterpillars and Sitka willows (Experiments I-IV) gave evidence for communication. Prior to these experiments, the study trees had been heavily attacked during 1976 and 1977 by tent caterpillars during a natural outbreak. On the other hand, the trees had not experienced attack by fall webworms for at least three years prior to our experiments with webworms, and probably not during their lifetime since they were only five years old in 1980, and no outbreaks of fall webworms occurred in the area from 1975-1980. Interestingly, in 1980, the first year of webworm attack (Fig. 5), a response was seen only in the attacked trees, whereas during the second and third years of attack (Figs. 6 and 7), evidence was obtained for a response in both attacked and unattacked trees. Conceivably, communication between attacked and unattacked willows requires prior "immunization" of the site by the attacking insects, even though the unattacked trees receiving and responding to the signals do not themselves need to have been previously attacked.

The results of Experiments I-V were presented at the Annual Meeting of the American Chemical Society in March 1982. Since no evidence could be found for a below-ground pathway of information transfer between the trees, airborne communication seemed likely. Also, trees and other plants commonly emit traces of organic chemicals of unknown function into the atmosphere, and the amounts and types of these emissions can be altered by plant damage (see below), suggesting that plants may be sensitive to airborne pheromonal signals produced by nearby damaged plants.[6,18]

Supporting Evidence From Other Workers

Following the American Chemical Society Conference, Baldwin and Schultz initiated laboratory experiments to detect possible changes in the defensive chemistry of undamaged plants near damaged ones.[23] They confined individually potted sugar maple seedlings in two growth chambers. They then tore leaves on some of the plants in one chamber and compared the phenolic chemistry of the damaged plants and undamaged plants (communication controls) in the same chamber with the undamaged plants in the separate chamber (true controls), for several days following damage. Increased levels of leaf total phenolics and tannins were found in leaves of the damaged plants and in the communication controls compared to the true controls. Similar results were obtained with poplar. Their results[23] are consistent with the field results from willow and support the proposed pheromonal mechanism.[6,18] Perry and Pitman found changes in foliage toxicity of Douglas-fir to western spruce budworm that occurred coincidentally with the appearance of a rapidly-building natural population of budworm in the vicinity of their study trees.[24] They suggest inter-tree communication as a possible explanation for this result.

Plant Volatiles

Volatile terpenes, non-terpene hydrocarbons and their derivatives emitted by ripening fruits and flowers function as attractants to fruit dispersers and pollinators.[25-27] However, the amount of volatile organic compounds emitted by these plant reproductive structures is dwarfed by that emitted from foliage. Most plants, especially trees, emit low concentrations of ethylene,[28-30] terpene hydrocarbons[31-34] and other volatile organic compounds[20,21,35,36] into the atmosphere from their leaves. Ethylene is a plant hormone. It plays important roles in fruit maturation, leaf senescence, and other processes,[28,29] but the function of other volatile organic foliage emissions is unknown.

Much work on plant volatiles has centered on monoter-penes and isoprenes (hemiterpenes) emitted by forest trees and shrubs because these emissions are often the major component of atmospheric hydrocarbon "pollutants" near forested areas.[34] Their photo-oxidation products are thought to be the source of atmospheric hazes often observed over forests.[37,38] Rasmussen estimated that the world's forests release approximately 175 X 10^6 tons of hydrocarbons, mainly foliage terpenes, per year into the atmosphere.[20] This is roughly six times the amount of hydrocarbons released by human activities.[34] The monoterpenes α-pinene, camphene, β-pinene, limonene, myrcene, and β-phellandrene are the primary volatiles emitted by conifers. The leaves of angiosperm trees and shrubs also commonly emit volatiles, including α-pinene, and especially isoprene, but most of their emis-sions are non-terpenoid and remain unidentified.[20,21] Many angiosperms are known to emit hexenal, a non-terpene.[21-35]

Interestingly, damage or stress of plants can lead to increases and qualitative changes in plant emissions. A variety of physical and biotic stresses of plants, including drought, freezing, disease, and mechanical damage can lead to increased ethylene emission.[28,29,39,40] Rose leaves infested with spider mites release more ethylene than controls.[41,42] Similarly ethylene production by Pinus radiata is stimulated by attack from the wood wasp Sirex and its associated fungus.[43] Trans-2-hexenal, which has been extracted from the leaves of many plant species, appears to be released mainly in response to damage of the leaves and is often not detectable in the atmosphere surrounding intact leaves or in extracts of leaves prepared in the absence of oxygen.[21,35,44] It is thought to be responsible for the characteristic smell of freshly mown grass.[44] Crushed pine needles have been reported to release ethane.[36]

Not surprisingly, chopped foliage of Douglas-fir emits far greater quantities of terpenes than intact foliage, due to disruption of resin ducts in the needles.[45] However, even gentle handling of foliage can cause large quantitative and qualitative changes in terpene emissions. Only isoprene was detected over undisturbed Eucalyptus camaldulensis foliage, but the same foliage disturbed by handling when the plant was placed in a study chamber emitted isoprene, α-pinene, cineole, limonene, α-terpinene, and ρ-cymene.[21] Isoprene was also the major volatile emitted by unstressed Salix babylonica foliage, but under transpirational stress due to confinement in an unvented chamber, the foliage emitted isoprene, unidentified alcohols and other compounds.[2]

Exposure to ethylene can lead to changes in the secondary chemistry and disease resistance of plants. Ethylene is thus a prime candidate as a messenger substance for communication between plants in response to damage. Activation of PAL/CAH enzyme systems (involved in phenol synthesis) and increases in levels of phenoloxidase enzymes (implicated in phenol-mediated disease[46,47] and insect[48] resistance) have been observed on exposure of plants to ethylene.[30,49] Ethylene treatment of bean leaves gives a 50-fold increase in chitinase activity.[50] The function of this enzyme in beans is unknown, though Abeles and coworkers suggested that it may have an antibiotic function since green plants do not contain chitin while insects and fungi do.[30,50] Pisatin, isocoumarin, and chlorogenic acid increase in concentration in peas, carrots, and other plants on exposure to ethylene, but, on the other hand, decreases in phenolic levels in some plants have also been seen.[30] Similarly, both increases and decreases in disease resistance have been observed in plants exposed to ethylene.[30,50] Chrominski and coworkers[51] report decreases and increases in the rate of nymphal development and decreases in adult longevity of grasshoppers feeding on rye exposed to ethylene. The effects depended on duration of exposure. However, in these studies, it is unclear whether ethylene acted directly on the insects (as the authors assume) or altered the nutritional quality of the plants. Whether other volatiles such as terpenes or hexenal released from foliage can cause changes in chemistry or resistance of plants is unknown.

There are a number of other pertinent questions. Are changes in emitted volatiles, caused by plant damage, confined to those emitted from the damaged tissue or do changes also occur in volatiles emitted from undamaged foliage on damaged plants? The former would suggest passive release by the plant, whereas the latter would suggest active release, and it could lead to amplification of the signal. Do volatiles released from damaged tissue trigger inducible defense in other undamaged portions of the same plant, or are systemic defensive responses in plants largely controlled by internal messengers? Is inter-plant communication involved in other synchronized behaviors of plants such as leaf flush, leaf drop, mass flowering and seed masting? Do plants communicate via root grafts[52,53] or mycorrhizal root connections?[54,55]

CONCLUSION

The evidence suggesting that information can be trans-ferred from damaged plants to nearby undamaged ones is limited but growing. Experiments with willows damaged by insects in the field, provided the first evidence for such communication effects.[6,18] The laboratory experiments of Baldwin and Schultz with maples and poplars, initiated as a direct follow-up of the willow work, provide strong supporting evidence.[23] The results of Perry and Pitman with Douglas-fir[24] are also consistent with communication between plants. In experiments with willows, below-ground communication was unlikely because no macroscopic connections were found with the roots of neighboring plants; the distances over which the effects occurred were large (up to 100 m, through intervening trees); and a trench between damaged and undamaged plants did not

eliminate the effect. In the experiments of Baldwin and
Schultz with maples and poplars, below-ground communication
was impossible because they used individually potted plants.
It therefore seems likely that an airborne vehicle of information transfer is involved. This hypothesis is reasonable
considering that plants commonly emit trace quantities of
volatile organics in amounts and types that vary with the
degree of damage sustained by the plants. One of these
volatiles, ethylene, is known to exert important physiological effects on exposed plant tissues. However, it should
be emphasized that the proposed airborne pheromonal mechanism
of communication is hypothetical and will remain so until
active volatiles have been isolated which can induce decreased
nutritional quality in exposed plants. In addition, below-
ground communication is possible in other cases. Grafts and
mycorrhizal connections between the roots of adjacent plants
are common, and many substances are known to be exchanged
between plants via these pathways.

The implications of these preliminary results for the
theory of plant-herbivore interactions, herbivore population
dynamics, and pest-control in forestry and agriculture are
substantial. If corroborated, the existence of pheromonal
communication between plants will require the restructuring
of our philosophy of relationships among plants and herbivores.

ACKNOWLEDGMENTS

I thank G.H. Orians for encouragement, L. Erckmann for
carrying out much of the experimental work, and A.B. Adams,
C.J. Baron, J.C. Bergdahl, and R. Hagen who also participated. I thank Mr. R.H. DeBoer, Stoneway, Inc., Mr. W.D.
Robertson, Mr. R.A. Marr, and the Seattle-Tacoma International Airport Authorities for use of their facilities, and
Dr. J.P. Donahue for identifying the insects. This research
was supported by National Science Foundation grants DEB
77-03258 and DEB 80-05528 to G.H. Orians and D.F. Rhoades.

REFERENCES

1. WALLACE, J.W., R.L. MANSELL, eds. 1976. Biochemical
    interaction between plants and insects. Rec. Adv.
    Phytochem. 10, Plenum Press, New York, 425 pp.

2. HARBORNE, J.B. 1978. Biochemical aspects of plant and animal coevolution, Academic Press, New York, 435 pp.
3. ROSENTHAL, G.A., D.H. JANZEN, eds. 1979. Herbivores: their interaction with plant secondary metabolites, Academic Press, New York, 718 pp.
4. MATTSON, W.J., N.D. ADDY. 1975. Phytophagous insects as regulators of forest primary production. Science 190: 515-522.
5. RHOADES, D.F. 1979. In G.A. Rosenthal, D.H. Janzen, eds., op. cit. Reference 3, pp. 3-54.
6. RHOADES, D.F. 1983. In Variable plants and herbivores in natural and managed systems. (R.F. Denno, M.S. McClure, eds.), Academic Press, New York, pp. 155-220.
7. WHITE, T.C.R. 1969. An index to measure weather-induced stress of trees associated with outbreaks of psyllids in Australia. Ecology 50: 905-909.
8. WHITE, T.C.R. 1974. A hypothesis to explain outbreaks of looper caterpillars with special reference to populations of Selidosema suavis in a plantation of Pinus radiata in New Zealand. Oecologia 16: 279-301.
9. WHITE, T.C.R. 1976. Weather, food, and plagues of locusts. Oecologia 22: 119-134.
10. BENZ, G. 1977. Eucarpia/IOBC working group breeding for resistance to insects and mites. Bull. SROP 1977/3. (Report from 1st meeting held at Wageningen, The Netherlands, 7-9 Dec. 1976, pp. 155-159).
11. HAUKIOJA, E. 1980. On the role of plant defenses in the fluctuation of herbivore populations. Oikos 35: 202-213.
12. BALTENSWEILER, W. 1968. In Insect abundance. (T.R.E. Southwood, ed.), Blackwell, Oxford, pp. 88-97.
13. FYE, R.E. 1974. In Proceedings of the summer institute on biological control of plant insects and diseases. (F.G. Maxwell, F.A. Harris, eds.), Univ. Miss. Press, Jackson, pp. 46-61.
14. STEHR, F.W. 1974. In F.G. Maxwell, F.A. Harris, eds., op. cit. Reference 13, pp. 124-136.
15. WATERS, W.E., R.W. STARK. 1980. Forest pest management: concept and reality. Annu. Rev. Entomol. 25: 479-509.
16. HAUKIOJA, E., T. HAKALA. 1975. Herbivore cycles and periodic outbreaks. Formulation of a general hypothesis. Rep. Kevo Subarctic Res. Stat. 12: 1-9.

17. RHOADES, D.F. 1984. Offensive-defensive interactions
    between herbivores and plants: their relevance in
    herbivore population dynamics and ecological theory.
    Am. Nat. (in press).
18. RHOADES, D.F. 1983. Responses of alder and willow to
    attack by tent caterpillars and webworms: evidence
    for pheromonal sensitivity of willows. Am. Chem.
    Soc. Symp. Ser. 208: 55-68.
19. SWAIN, T.S., W.E. HILLIS. 1959. The phenolic constit-
    uents of Prunus domestica. I. The quantitative
    analysis of phenol constituents. J. Sci. Food
    Agric. 10: 63-68.
20. RASMUSSEN, R.A. 1972. What do hydrocarbons from trees
    contribute to air pollution? J. Air Poll. Contr.
    Assoc. 22: 537-543.
21. RASMUSSEN, R.A. 1970. Isoprene: identified as a
    forest-type emission to the atmosphere. Env. Sci.
    Technol. 8: 667-671.
22. WALDBAUER, G.P. 1968. In Advances in insect physiology.
    (J.W.L. Beament, J.E. Treherne, V.B. Wigglesworth,
    eds.), Vol. 5, Academic Press, New York, pp. 229-288.
23. BALDWIN, I.T., J.C. SCHULTZ. 1983. Rapid changes in
    tree leaf chemistry induced by damage: evidence for
    communication between plants. Science 221: 277-279.
24. PERRY, D.A., G.B. PITMAN. 1983. Genetic and environ-
    mental influence in host resistance to herbivory:
    Douglas-fir and the western spruce budworm. Z. Ang.
    Ent. 96: 217-228.
25. DODSON, C.H. 1970. The role of chemical attractants
    in orchid pollination. Annu. Biol. Colloquiam,
    Corvallis, Oregon 1968 29: 83-107.
26. BERGSTROM, G. 1978. In J.B. Harborne, ed., op. cit.
    Reference 2, pp. 207-231.
27. NURSTEIN, H.E. 1970. In The biochemistry of fruits
    and their products. (A.C. Hulme, ed.), Academic
    Press, New York, pp. 239-268.
28. BURG, S.P. 1965. The physiology of ethylene formation.
    Annu. Rev. Plant Physiol. 13: 265-302.
29. PRATT, H.K., J.D. GOESCHL. 1969. Physiological roles
    of ethylene in plants. Annu. Rev. Plant Physiol.
    17: 541-584.
30. ABELES, F.B. 1973. Ethylene in plant biology. Academi
    Press, New York, 302 pp.
31. HANOVER, J.W. 1972. Factors affecting the release of
    volatile chemicals by forest trees. Mitt. First
    Bundes-Versuch. Wien 97: 625-645.

32. ROBINSON, E., R.A. RASMUSSEN, H.H. WESTBERG, M.W. HOLDREN. 1973. Non-urban, non-methane, low molecular weight hydrocarbon concentrations related to air mass identification. J. Geophys. Res. 78: 5345-5351.
33. FREELAND, W.J. 1980. Insect flight times and atmospheric hydrocarbons. Am. Nat. 116: 736-742.
34. SMITH, W.H. 1981. Air pollution and forests. Springer-Verlag, New York, 379 pp.
35. MAJOR, R.T., P. MARCHINI, A.J. BOULTON. 1963. Observations on the production of α-hexenal by leaves of certain plants. J. Biol. Chem. 238: 1813-1816.
36. SUMIMOTO, M., M. SHIRAGA, T. KONDO. 1975. Ethane in pine needles preventing the feeding of the beetle Monochamus alternatus. J. Insect Physiol. 21: 713-722.
37. WENT, F.W. 1960. Blue hazes in the atmosphere. Nature 187: 641-643.
38. WESTBERG, H.H., R.A. RASMUSSEN. 1972. Atmosphere phytochemical reactivity of monoterpene hydrocarbons. Chemosphere 4: 163-168.
39. YANG, S.F., H.K. PRATT. 1968. In Biochemistry of wounded plant tissues. (G. Kahl, ed.), Walter de Gruyter, New York, pp. 595-622.
40. BOLLER, T., H. KENDE. 1980. Regulation of wound ethylene in plants. Nature 286: 259-264.
41. WILLIAMSON, C.E. 1950. Ethylene, a metabolic product of diseased or injured plants. Phytopathology 40: 205-208.
42. WILLIAMSON, C.E., A.W. DIMMOCK. 1953. Ethylene from diseased plants. Yearbook of Agriculture (USDA), pp. 881-886.
43. SHAIN, L., W.E. HILLIS. 1972. Ethylene production in Pinus radiata in response to Sirex-Amylostereum attack. Phytopathology 62: 1407-1409.
44. NYE, W., H.A. SPOEHR. 1943. The isolation of hexenal from leaves. Arch. Biochem. 2: 23-35.
45. RADWAN, M.A., W.D. ELLIS. 1975. Clonal variation in monoterpene hydrocarbon vapors of Douglas-fir foliage. For. Sci. 21: 63-67.
46. LEVIN, D.A. 1971. Plant phenolics: an ecological perspective. Am. Nat. 105: 157-181.
47. KIRALY, Z. 1980. In Plant disease. (J.G. Horsfall, E.B. Cowling, eds.), Vol. 5, Academic Press, New York, pp. 201-224.

48. RHOADES, D.F. 1977. In Creosote bush. (T.J. Mabry,
     J.H. Hunziker, D.R. DiFeo, eds.), Dowden, Hutchinson,
     and Ross, Stroudsberg, PA, pp. 135-175.
49. CHRISTOFFERSON, R.E., G.C. LATIES. 1982. Ethylene
     regulation of gene expression in carrots. Proc.
     Nat. Acad. Sci. (USA) 79: 4060-4063.
50. ABELES, F.B., R.P. BOSSHART, L.E. FORRENCE, W.H. HABIG.
     1971. Preparation and purification of glucanase and
     chitinase from bean leaves. Plant Physiol. 47: 129-
     134.
51. CHROMINSKI, A., S. NEUMANN VISSCHER, R. JERENKA. 1982.
     Exposure to ethylene changes nymphal growth rate and
     female longevity in the grasshopper Melanopus
     sanguinipes. Naturwissenschaften 69: (in press).
52. GRAHAM, B.F. JR., F.H. BORMANN. 1966. Natural root
     grafts. Bot. Rev. 32: 255-292.
53. EPSTEIN, A.H. 1978. Root graft transmission of tree
     pathogens. Annu. Rev. Phytopathol. 16: 181-192.
54. ATSATT, P.R. 1970. Biochemical bridges between vascular
     plants. Annu. Biol. Colloquiam, Corvallis, Oregon
     1968 29: 53-68.
55. HARLEY, J.L., S.E. SMITH. 1983. Mycorrhizal symbiosis,
     Academic Press, New York.

Chapter Nine

ADAPTATION TO RESOURCE AVAILABILITY AS A DETERMINANT OF
CHEMICAL DEFENSE STRATEGIES IN WOODY PLANTS

J.P. BRYANT AND F.S. CHAPIN III

Institute of Arctic Biology
University of Alaska
Fairbanks, Alaska 99701

P. REICHARDT AND T. CLAUSEN

Department of Chemistry
University of Alaska
Fairbanks, Alaska 99701

INTRODUCTION

Browsing mammals are usually generalist herbivores.
For example, in boreal forests snowshoe hares (Lepus americanus) and mountain hares (L. timidus) usually require a
multi-species diet in order to survive during winter.[1,2]

Two hypotheses attempt to explain the generalist feeding
behavior of browsing mammals: (a) Browsing mammals must
feed upon several plant species in order to avoid ingestion
of an overdose of one plant's toxic secondary metabolites.[2,3]
(b) Browsing mammals must feed upon several plant species to
optimize nutrient or energy intake.[4-8] Here we test these
two hypotheses by considering the phytochemical basis for
winter food selection by snowshoe hares in an Alaskan taiga
forest. We conclude by discussing our findings in light of
presently held views of plant-herbivore interactions.

WINTER BROWSING BY MAMMALS IN BOREAL FORESTS: AN OVERVIEW

     In winter, boreal forest browsing mammals such as hares
feed upon a variety of woody plant growth forms (evergreen
versus deciduous), species, growth stages (adult versus
juvenile) and parts (e.g. foliar buds, staminate catkins,
internodes of small twigs and bark of larger stems). Given
this potential dietary variety, boreal forest browsing mammals have very similar winter food preferences. Slowly
growing evergreens, such as black spruce (Picea mariana) and
Labrador tea (Ledum groenlandicum and L. palustre), and
nitrogen-fixing trees and shrubs like alders (Alnus) and
Sheperdia canadensis, are low preference winter foods.
Rapidly growing, early successional deciduous trees and
shrubs, such as many willows (Salix), poplars (Populus) and
birches (Betula), are the preferred winter foods.[1,2,9-15]
Within each species twigs of juvenile-form individuals are
less preferred as winter food than twigs of adult-form
individuals.[2,10,12,14,16-27]

     Three observations indicate that this pattern of food
selection is largely a consequence of avoidance of repellent
secondary plant metabolites. (a) Avoidance of nitrogen fixing species such as the alders argues against selection for
an "optimal mixture of nutrients". On the basis of their
nutrient content, nitrogen-fixing trees and shrubs should be
"optimal food", because they contain more of the nutrients
required by browsing mammals than preferred woody plant
species.[12] (b) Avoidance of late successional evergreens
such as black spruce indicates that food selection is not
based upon "energetic optimization". Winter dormant twigs
of late successional evergreens such as black spruce contain
higher concentrations of lipids and easily digestible sources
of energy, carbohydrates, than preferred woody plant species

(Bryant, unpub.).[12] (c) The low preference of juvenile form
woody plants is not caused by low nutrient concentration.
On the basis of mineral nutrients juvenile- and mature-form
plants appear identical.[2,10,12,14,18-20,22,23,25,27]
Moreover, snowshoe hares and mountain hares often reject
plant parts that contain high concentrations of mineral
nutrients, lipids and carbohydrate (winter-dormant foliar
buds, staminate catkins, and 0-1.5 mm diameter internodes of
some willows, poplars, birches and alders) and eat apparently
less "nutritious" larger diameter (1.5-4 mm) woody internodes.[2,12,20,22,24,25,28] This selection of plant parts
argues strongly against the optimal foraging hypothesis.

Because of the lack of correlation between plant nutrient
or digestible energy concentrations and winter food selection
by boreal forest vertebrate herbivores, we initiated a project
to determine if plant secondary metabolites influence winter
food selection by snowshoe hares. Two of the plant species
we have studied in detail are Alaska paper birch (B.
resinifera = B. papyrifera ssp. humilis)[29] and green alder
(A. crispa).[30] These species were chosen because (a) they
range from rapidly growing early successional (B. resinifera)
to more slowly growing mid to late successional (A. crispa)
in habit;[14,30] (b) snowshoe hare preference for the adult-
growth-stage as compared to juvenile-growth-stage ranges from
extremely pronounced (B. resinifera) to slightly pronounced
(A. crispa);[12,14] and (c) hares feed selectively upon some
parts of these plants in winter, preferring 1.5-4 mm diameter
winter-dormant internodes and rejecting smaller internodes
(0-1.5 mm diameter), foliar buds and staminate
catkins.[12,14,22,25] Thus, hare use of these species covers
the range of hare use of winter-dormant woody plants in
general.

PHYTOCHEMISTRY OF WINTER-DORMANT ALASKA PAPER BIRCH AND
GREEN ALDER

Initial efforts to identify chemical factors which
influence the use of winter-dormant birch and alder by hares
have established that low palatability growth stages (juvenile) and parts (buds, staminate catkins and 0-1.5 mm
diameter internodes) are often characterized by ether
soluble feeding deterrents[12,18,19,25,28] that are toxic to
cervid rumen microbes[31] (and Risenhoover, unpub.) and may
be toxic to hares.[2,13,20,22,25] We have attempted to

elucidate the molecular basis for these observations. The goals of our research have been to (a) identify individual components of the ether extractable fraction of these plants; (b) determine by bioassay which of these components deter snowshoe hare feeding; (c) quantify the levels of active substances in growth stages and parts; and (d) relate intraplant allocation of deterrent substances to snowshoe hare use of these plants.

Chemical Methods and Findings

From the ether extract of green alder we have isolated and identified four major components: pinosylvin (I), pinosylvin monomethyl ether (II),[22] pinostrobin (IV), and β-phenethyl cinnamate (V) (Clausen, unpub. data). On the other hand, the ether extractables of Alaska paper birch (at least in its juvenile growth stage) are dominated by a single substance, papyriferic acid (VI) (Fig. 1; Table 1).[25,32]

Table 1. Representative analyses for secondary metabolites of Alaska paper birch (B. resinifera) and green alder (A. crispa).

Growth Form and Part	Pinosyl-vin[1,A]	Pinosylvin[2,A] Monomethyl Ether	Pinostro-bin[2,A]	β-Phenethyl[A] Cinnamate	Papyri-feric[3,B] Acid
		Concentration (% Dry Wt.)			
Mature					
Bud	3.40	2.60[a]	2.50[a]	2.90	1.50[a]
Catkin	0.90[a]	1.70[a]	1.50[a,b]	1.10[a]	2.90[a]
Current year internode[4]	0.04	0.05[b]	0.02[c]	0.02[b]	0.45[a]
Juvenile					
Bud	1.30[a]	1.40[a]	1.00[b]	0.70[a]	0.79[a]
Current year internode[4]	0.11	0.06[b]	0.04[c]	0.02[b]	11.30

Quantification by (1) HPLC, (2) GLC, (3) ^{1}H-NMR; (4) 0-4 mm diameter.

a,b,c Values within a column with same superscript do not differ at P<0.05.

A = Alder metabolite; B = Birch metabolite.

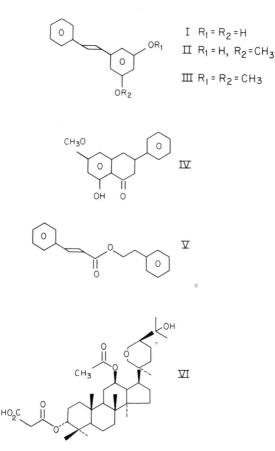

$$I \quad R_1 = R_2 = H$$
$$II \quad R_1 = H, \; R_2 = CH_3$$
$$III \quad R_1 = R_2 = CH_3$$

Fig. 1. Structures of secondary metabolites in Alaska paper birch (B. resinifera) and green alder (A. crispa): pinosylvin (I), pinosylvin monomethyl ether (II), pinosylvin dimethyl ether (III), pinostrobin (IV), β-phenethyl cinnamate (V), papyriferic acid (VI).

The abilities of these compounds to deter feeding by snowshoe hares were analyzed by a bioassay procedure described in detail elsewhere.[22,25] In short, hares were offered oatmeal treated with carrier solvents (control) as well as food treated with solutions of test compounds. Deterrent potency is defined as a preference index (PI) which is the ratio of the fraction of adulterated food consumed to the fraction of the control consumed during the

Table 2. Effect of secondary metabolites from Alaskan paper
birch (B. resinifera) and green alder (A. crispa) on snowshoe
hare feeding preferences.

Compound	% Compound Added to Oatmeal (Dry Wt. Basis)	Number of Hares	($\bar{x} \pm S\bar{x}$)
Papyriferic Acid (VI)	2.0	5	0.28 ± 0.09
Pinosylvin (I)	1.5	6	0.05 ± 0.06
Pinosylvin Monomethyl Ether (II)	1.5	6	0.02 ± 0.02
β-Phenethyl Cinnamate (V)	2.0	7	0.42 ± 0.17
Pinostrobin (IV)	1.5	6	0.63 ± 0.17

* PI = Preference Index = % treated oatmeal eaten ÷ % control
oatmeal eaten.

Data are from Clausen, unpub. and Reichardt, unpub.

time of the feeding trial. Results for the compounds pertinent to the present discussion are presented in Table 2.
From these data the following points are clear: (a) deterrent potencies vary widely for individual compounds with some
having virtually no activity; (b) deterrent compounds can
have a wide variety of structural features; and (c) in some
cases small differences in structure cause large differences
in activity. For example, both pinosylvin (I) and pinosylvin
monomethyl ether (II) extracted from green alder are highly
deterrent to both snowshoe hares (Table 2) and microtine
rodents (Henttonen, unpub.), but pinosylvin dimethyl ether
(III), which does not occur in green alder, is not particularly repellent to snowshoe hares (PI = 0.63 ± 0.09; $\bar{x} \pm S\bar{x}$;
1.5% loading). Thus it is clear that considerations of
chemically-based plant defenses against hares must rely upon
levels and activities of individual substances rather than
gross fractions such as total phenols, tannins or resins.

Having established that both Alaskan paper birch and
green alder contain substances that deter feeding by snowshoe hares, we then directed our attention toward quantifying
these compounds in parts of these plants. These studies
addressed two questions about hare feeding behavior: Can
levels of deterrent compounds explain the hares' preference

for mature over juvenile growth stage plants? Can levels of these compounds explain the hares' rejection of plant parts within growth stages?

In the case of Alaska paper birch, it is clear that papyriferic acid (VI) plays a defensive role in the juvenile growth stage (Table 2).[25] Not only is the concentration of VI quite high in the juvenile material; it is largely concentrated in deposits on the <u>exterior</u> of current years' twigs.[25,29,32] On the other hand, the hares' preference for mature over juvenile phase alder does not appear to be caused by the levels of any single substance, but can be accounted for by the combined concentrations of I and II (Table 2; Clausen, unpub.). Furthermore, neither the hares' preference for mature over juvenile material nor the difference in levels of defensive substances in alder internodes is as dramatic as those in birch. Thus plasticity of allocation of defensive substances to internodes during ontogeny is less in the slowly growing alder as compared to the more rapidly growing birch.

The hares' rejection of buds and catkins from alder can, again, be explained by consideration of the levels of I and II (Table 2; Clausen, unpub.), although in this case one needs not consider the levels of both compounds in order to rationalize the feeding behavior.[22] In birch it is unlikely that the concentrations of VI found in buds and catkins are high enough to account for their rejection (Table 2). However, in this case the deterrent effect of VI is supplemented by an unpalatable (and presently uncharacterized) volatile fraction.[25] While we have not explicitly addressed the question of why 0-1.5 mm internodes are of low palatability to hares, the allocation of substances such as VI to the surface of bark indicates that rejection may be a consequence of increased secondary metabolite concentration resulting from an increased bark/wood ratio in these small twigs.

WHY DO BOREAL FOREST BROWSING ANIMALS REQUIRE A MULTISPECIES DIET IN WINTER?

While our results demonstrate how specific secondary plant metabolites control snowshoe hare use of Alaska paper birch and green alder, they can also be viewed as an example of how secondary plant metabolites influence boreal browsing mammals' use of winter-dormant woody plants in general. As

is the case for most boreal forest browsing mammals,[12] the
snowshoe hare is an obligate generalist herbivore in winter
(Bryant, unpub.).[1,15] Feeding experiments have demonstrated
that when fed winter-dormant twigs from the height range
normally available to hares in winter (0-0.5 m above the
snow surface), the snowshoe hare usually requires at least
two highly palatable browse species in its diet to meet its
daily energy requirements (Bryant, unpub.).[1,33] Moreover,
as the palatability of dietary components declines, a
greater diversity of plant species, growth stages and parts
is required in the winter diet in order for the snowshoe
hare to maintain weight.[1] Similar results have been obtained
for the snowshoe hare's Eurasian ecological equivalent (L.
timidus)[2,13,20] and Alaskan moose (Alces alces gigas).[34]

    Efforts to explain generalist foraging by large
mammalian herbivores usually invoke nutritional or energetic
optimization as the cause. Mammals allegedly select a
varied diet in order to optimize intake of nutrients or
energy.[4-8] Our present results and results of several
other recent studies do not support this
generalization.[2,10,12-14,18-20,22,25,28,35-38] Rather,
increasing evidence now indicates that boreal forest browsing mammals require a multispecies diet in winter, because
almost all winter-dormant woody plants in boreal forests
contain sufficient quantities of toxic secondary metabolites
to be poisonous if eaten in large enough quantities to meet
the herbivore's daily metabolizable energy
requirements.[2,13,14,20,22,25,36-38]

    While we do not yet have direct evidence for this
hypothesis, our results and those of others provide
circumstantial support. When fed diets of highly unpalatable woody plant species, growth stages or parts, snowshoe
hares, mountain hares and moose exhibit symptoms of a toxin
overdose; i.e., refusal to eat after first eating some of
the browse in question (Bryant, unpub.),[13,25] lethargy,
reduced gut mobility,[34] excessive excretion of electrolytes
(Na),[2,13,20,25] excessive excretion of phenols and glucuro-
nides,[23,36-38] and frequently death in 2-4 days.[1,25] For
example, when 0-1.5 mm winter-dormant twigs of mountain
birch (B. pendula), winter-dormant twigs of red heather
(Calluna vulgaris) or Scots pine (Pinus sylvestris) needles
that are low preference foods of Fennoscandian mountain
hares are fed to caged hares, the animals first eat some
of the browse in question and then stop eating entirely.[13]

Concommitant with ingestion of these plant tissues is a
sharp increase in the excretion of sodium[2,13,20] and
phenolics[36-38] via the urine as compared to sodium excretion
when feeding upon a browse diet of sufficient palatability
and food value (Salix caprea) that these hares can maintain
weight in winter.[2] Similar results are obtained when snowshoe hares are fed 0-1.5 mm twigs of winter-dormant Alaska
paper birch.[25] The diethyl ether extract of red alder
(A. rubra) has proven extremely toxic to deer rumen
microbes.[31] Experiments have verified that pinosylvin (I),
pinosylvin monomethyl ether (II) and papyriferic acid (VI)
are toxic to laboratory mice at low concentrations.[25,39]
In the case of pinosylvin and pinosylvin methyl ether, the
toxicity appears to lie in uncoupling of oxidative phos-
phorylation.[40] In a separate experiment both papyriferic
acid and the steam distillate of juvenile birch proved
extremely toxic to elk (Cervis canadensis) rumen microbes
(Risenhoover, unpub.) at dosages below those found in
juvenile Alaska paper birch winter dormant twigs.

In summary, we have presented a variety of evidence
indicating that individual toxic secondary plant metabolites
strongly influence food choice by boreal forest browsing
mammals. We further argued that, as predicted by Freeland
and Janzen[3] and Pehrson[2,13,20] the dietary generalism
exhibited by these herbivores is a consequence of avoiding
ingestion of large quantities of toxic secondary plant
metabolites. We conclude by providing a theoretical frame-
work which can account for allocation of resources by boreal
woody plants to defense against winter browsing by vertebrate
herbivores.

EVOLUTION OF CHEMICAL DEFENSES BY BOREAL WOODY PLANTS

Resource Availability as a Determinant of Defensive Evolution

    Plants occurring on infertile soils (associated with
late successional stages in boreal regions) or in shade
generally cannot acquire sufficient resources to support
rapid growth. The evolutionary response to resource limitation has been an inherently slow growth rate. Such plants
grow slowly even in the most favorable environments and
have low capacities to photosynthesize and absorb
nutrients.[41,42] Although growth is slow, plants and plant
parts tend to be long-lived. Thus long-lived evergreen

species are commonly found in the most nutrient-deficient
or shaded sites.[42-44] Slow turnover of plant parts is
advantageous in a low-nutrient environment, because every
time a plant part is shed, it carries with it approximately
half its nitrogen and phosphorus pool;[42] such nutrients are
not readily replaced.[14] However, greater leaf longevity
carries with it certain disadvantages. First, a long-lived
leaf is more likely to encounter unfavorable physical
conditions. The high fiber content, low water content and
thick cuticle of many evergreen leaves may be in part an
evolutionary response to unfavorable conditions such as
winter desiccation or summer drought.[45] Such characteristics
also lower the palatability[44] and digestibility of forage
to herbivores[46,47] and thus may also be an evolutionary
response to herbivory.[14] Secondly, long-lived leaves are
available for attack by herbivores and pathogens for a
longer time than short-lived leaves and therefore have a
greater probability of being attacked before being
shed.[48-50] The obvious evolutionary response in slowly
growing evergreens has been the production of secondary
chemicals that deter and are often toxic to
herbivores.[12,14,44,48-50]

Evergreen species also differ from deciduous species in
their lack of major storage organs in stems or roots. Thus
leaves and twigs comprise a larger proportion of total
biomass in evergreen than in deciduous species and are the
source of stored nutrients to support growth.[51,52] This
resource allocation strategy greatly limits their capacity
to replace destroyed above-ground parts through compensatory
growth.[51-53] Consequently these plants have a low physiological capacity to replace tissues eaten by herbivores and
are therefore expected to be strongly selected for chemical
defenses that deter herbivore attack.[12,14,44] Furthermore,
because allocation of carbon to growth in these species is
strongly limited by nutrients, their chemical defenses are
expected to contain no nitrogen, i.e. be derived from
shikimate and mevalonate.[14]

Species that have evolved in fertile soils on disturbed
sites have been selected to grow rapidly so as to outcompete
their neighbors and thus dominate available light and
mineral resources. Thus these species have been strongly
selected to allocate resources to growth at the expense of
defense. Adaptation to physical disturbance of above-ground
parts also entails selection for large below-ground carbon

and nutrient reserves that can be used in support of compensatory growth following disturbance.[51-53] In short, these
plants are preadapted to compensate for herbivory through
compensatory growth.[12,14,44,53] This preadaptation further
reduces the intensity of selection for antiherbivore
defenses.

Rapidly growing woody plants often have high leaf and
twig turnover rates. New leaves and twigs are produced at
the top of the canopy of adult plants where the light environment is more favorable than that of older leaves. The
shaded lower leaves and twigs then develop a negative carbon
balance and are shed.[51,52,54] Moreover, the inevitable
nutrient loss associated with rapid leaf and twig turnover
is not a strong selective influence on these plants because
nutrients are readily available in early successional sites
on fertile soils. Therefore, moderate pruning of inefficient
lower branches does not negatively affect growth of adult
plants.[51,52] Such pruning should not strongly select these
plants for antiherbivore defenses.

In summary, there are several reasons to expect slowly
growing, late successional woody plants to be strongly
selected for constitutive antiherbivore defenses. Their
long-lived leaves and twigs are expensive to replace if lost
to an herbivore because (a) the necessary nitrogen and
phosphorus are in short supply in a low-nutrient environment,
and rates of carbon fixation are low in a low light environment; and (b) leaves and twigs constitute the major storage
organs of evergreen plants, and reserves therefore cannot be
mobilized from other plant parts to replace leaves and twigs
eaten by herbivores. On the other hand, early successional
woody plants are primarily selected to allocate resources
to growth. This adaptation limits their ability to allocate
resources to defense. Adaptations that enable these species
to replace above-ground parts destroyed by physical forces
such as wildfire preadapt these species for the ability to
compensate for herbivory through growth. Moreover, moderate
pruning of lower branches in adult individuals is not detrimental to plant fitness. Thus these rapidly growing woody
plants are not expected to be strongly selected for constitutive antiherbivore defenses.

Age-Specific Selection for Defense Against Winter Browsing

Woody plants in boreal forests are at particularly high risk of vertebrate browsing during the juvenile growth phase.[55-62] Because the reproductive value of an organism is greatest near first reproduction[63,64] and because severe browsing delays reproductive maturity,[51,52,65,66] heavy browsing of juvenile-phase woody plants can be expected to have a strong negative impact upon woody plant fitness.[14,23] Consequently, strong selection for defense against browsing during the juvenile phase of all woody plants is expected on the basis of plant life history alone.[14,23]

Severe pruning, decapitation or girdling of juvenile-phase plants, as occurs during periods of high snowshoe hare numbers,[15,55-62] reduces their rate of vertical growth and therefore their ability to compete for canopy dominance.[51,52] Consequently severe winter browsing of juvenile individuals of rapidly growing woody plants can strongly reduce their fitness. Thus, in contrast to the adult phase, which is selected more strongly for growth than defense, juveniles of rapidly growing woody plants are expected to allocate resources to defense at the expense of growth because such an allocation of resources ultimately increases their competitive potential.[12] This selective pressure enhances juvenile-phase defenses resulting from life history traits.

In summary, the juvenile phase of all woody plants is expected to be selected for defenses against vertebrate browsing, but during the adult phase only slowly growing woody plants are expected to be strongly selected for anti-herbivore defenses. Consequently ontogenetic variation in antiherbivore defense will be greater in rapidly growing than in slowly growing woody plants.[12,14]

Cost of Chemical Defense

Current antiherbivore defense theory predicts that the energetic cost of producing large quantities of carbon-based defensive substances, (e.g. quantitative defenses such as tannins and resins) limits growth in late successional woody plants such as black spruce; allocation of energetically expensive carbon to defense must be paid for by reduced growth.[48-50] We propose an alternative explanation of the correlation between slow growth rate and enhanced allocation of carbon to antiherbivore defense by late successional woody plants. Carbon-based defenses are relatively "cheap" for plants growing in infertile soils to produce, because nutrient limitation of growth makes carbon a relatively "cheap" commodity.[14,42] Slow growth rate is the cause rather than the consequence of allocation of carbon to defense by plants growing on infertile soils.[12,14,44]

These alternative hypotheses imply different responses by plants to increased mineral nutrition. If allocation of carbon to chemical defense limits plant growth, then fertilization will not increase growth at the expense of carbon-based defense. On the other hand, if growth is limited by nutrient availability then fertilization with growth-limiting nutrients should increase growth at the expense of defense.[14]

Fertilization of boreal forest evergreen woody plants with growth-limiting nutrients results in both increased growth and increased browsing by vertebrate herbivores. For example, fertilization of pine (P. contorta, P. resinosa and P. sylvestris) with growth-limiting nutrients (N, P, and K) caused increased growth and increased browsing by hare, moose and squirrels.[67,70] That increased browsing is not consistently correlated with increased browse nutrient concentration, particularly nitrogen, implies that reduced secondary metabolite concentration is the cause of increased browsing.[67,68] This hypothesis is favored by the observation that fertilization of woody plants with growth-limiting nutrients leads to reduced concentrations of carbon-based secondary metabolites such as shikimates and terpenes.[71,72] These results indicate that the carbon cost of secondary metabolite production does not limit the growth rate of late successional woody plants in boreal forests.

CONCLUSIONS

We conclude that individual, toxic secondary plant metabolites strongly influence both the palatability and food value of winter-dormant woody plant tissues to browsing mammals in boreal forests. The available evidence indicates that plant physiological adaptation to the availability of resources in the physical environment (carbon/nutrient balance) strongly constrains plant defensive evolution. Woody plants adapted to low-resource environments have inherently slow growth rates that limit their capacity to

replace browsed tissue through compensatory growth. These
low-resource-adapted plants have responded to herbivore
attack by evolving chemical defenses that are well developed
throughout the life of the plant. In contrast, high-
resource-adapted plants are preadapted to compensate for
herbivory through increased growth. This ability to compensate for herbivory through growth is a consequence of
adaptation to physical disturbance in nutrient-rich environments such as early stages of succession following wildfire.
Thus these plants allocate resources to growth at the expense
of defense except when juvenile. The juvenile phase of all
boreal woody plants has been strongly selected for defense
against browsing by vertebrates in winter.

We further propose that the carbon cost of defense in
late successional woody plants growing on nutrient deficient
soils may be low because the availability of nutrients rather
than carbon limits growth. Thus enhanced constitutive
defense of late successional (apparent plants) is not the
cause of their slow growth rate. It is a consequence of an
intrinsically slow growth rate that is an adaptation to
limited availability of resources in the physical environment.

ACKNOWLEDGMENT

This work was supported by NSF grants DEB-8207170 and
DEB 7823919.

REFERENCES

1. BOOKHOUT, T.A. 1965. The snowshoe hare in Upper
    Michigan and its biology and feeding coactions with
    white-tailed deer. Mich. Dept. Conserv. Res. Develop.
    Rep. No. 38, 198 pp.
2. PEHRSON, A. 1981. Winter food consumption and digesti-
    bility in caged mountain hares. In Proc. 1st Inter-
    national Lagomorph Conference. (K. Meyers, C.D.
    MacInnes, eds.), Guelph Univ., Canada, pp. 732-742.
3. FREELAND, W.J., D.H. JANZEN. 1974. Strategies in
    herbivory by mammals: the role of plant secondary
    compounds. Am. Nat. 108: 269-289.
4. WESTOBY, M. 1974. An analysis of diet selection by
    large generalist herbivores. Am. Nat. 108: 290-304.

5. BELOVSKY, G.E. 1978. Diet optimization in a generalist herbivore: the moose. Theor. Pop. Biol. 14: 105-134.

6. BELOVSKY, G.E. 1981. Food selection by a generalist herbivore: the moose. Ecology 62: 1020-1030.

7. BELOVSKY, G.E. 1984. Snowshoe hare optimal foraging and its implications for population dynamics. Theor. Pop. Biol. (in press).

8. SINCLAIR, A.R.E., C.J. KREBS, J.N.M. SMITH. 1982. Diet quality and food limitation in herbivores: the case of the snowshoe hare. Can. J. Zool. 60: 889-897.

9. LINDLOF, B., E. LINSTROM, A. PEHRSON. 1974. On activity, habitat selection and diet of the mountain hare (Lepus timidus L.) in winter. Viltrevy 9: 27-41.

10. KLEIN, D.R. 1977. Winter food preferences of snowshoe hares (Lepus americanus). In Alaskan Proc. Int. Congr. Game Biol., 13th, Atlanta, pp. 266-275.

11. FOX, J.F. 1978. Forest fires and the snowshoe hare - Canada lynx cycle. Oecologia 31: 349-374.

12. BRYANT, J.P., P.J. KUROPAT. 1980. Selection of winter forage by subarctic browsing vertebrates: The role of plant chemistry. Annu. Rev. Ecol. Syst. 11: 261-285.

13. PEHRSON, A. 1984. Maximal winter browse intake in captive mountain hares. Finn. Game Res. (in press).

14. BRYANT, J.P., F.S. CHAPIN, III, D.R. KLEIN. 1983. Carbon/nutrient balance of boreal plants in relation to vertebrate herbivory. Oikos 40: 357-368.

15. KEITH, L.B., J.R. CARY, O.R. RONGSTAD, M.C. BRITTINGHAM. 1984. Demography and ecology of a snowshoe hare population decline. Wildl. Monogr. (in press).

16. DIMOCK, E.J., II. 1974. Animal resistant Douglas-fir. How likely and how soon? In Wildlife and forest management in the Pacific Northwest. (H.C. Black, ed.), Oregon State Univ., Corvallis, Oregon, pp. 95-101.

17. LIBBY, W.J., J.V. HOOD. 1976. Juvenility in hedged radiata pine. Acta Horticulturae 56: 91-98.

18. BRYANT, J.P. 1981. The regulation of snowshoe hare feeding behavior in winter by plant antiherbivore chemistry. In K. Meyers, C.D. MacInnes, eds., op. cit. Reference 2, pp. 720-731.

19. BRYANT, J.P. 1981. Phytochemical deterrence of snowshoe hare browsing by adventitious shoots of four Alaskan trees. Science 313: 889-890.

20. PEHRSON, A. 1983. Digestibility and retention of food components in caged mountain hares (Lepus timidus L.) during the winter. Holarctic Ecol. 6: 395-403.

21. MOSS, R., G.R. MILLER, S.E. ALLEN. 1972. Selection of heather by captive red grouse in relation to the age of the plant. J. Appl. Ecol. 9: 771-781.

22. BRYANT, J.P., G.D. WIELAND, P.B. REICHARDT, V.E. LEWIS, M.C. McCARTHY. 1983. Pinosylvin methyl ether deters snowshoe hare feeding on green alder. Science 222: 1023-1025.

23. BRYANT, J.P., G.D. WIELAND, T. CLAUSEN, P. KUROPAT. 1984. Interactions of snowshoe hares and feltleaf willow (Salix alaxensis) in Alaska. Ecology. (in press).

24. FOX, J.F., J.P. BRYANT. 1984. Instability of the snowshoe hare and woody plant interaction. Oecologia. (in press).

25. REICHARDT, P.B., J.P. BRYANT, T.P. CLAUSEN, G. WIELAND. 1984. Defense of winter-dormant Alaska paper birch against snowshoe hare. Oecologia. (in press).

26. RYALA, P. 1966. Riekon jakiirunan talvisesta kasviravinnon valinnasta ja puissa roukalilusta. Suomen Riista 19: 79-93.

27. SINCLAIR, A.R.E., J.N.M. SMITH. 1984. Do plant secondary compounds determine feeding preferences of snowshoe hares? Oecologia 61: 403-410.

28. BARIKMO, J. 1976. Harens utnyhelse au bjork som vinter fod. Inst. Nat. Cons. N.L.H. Norway, pp. 125.

29. Dugle, J.R. 1966. A taxonomic study of western Canadian species in the genus Betula. Can. J. Bot. 44: 929-1007.

30. VIERECK, L.A., E.J. LITTLE, JR. 1972. Alaska Trees and Shrubs. U.S. Dept. Agric. Handb. No. 410, 265 pp.

31. RADWAN, M.A., G.L. CROUCH. 1974. Plant characteristics related to feeding preference by black-tailed deer. J. Wildl. Manag. 38: 32-41.

32. REICHARDT, P.B. 1981. Papyriferic acid: a triterpenoid from Alaskan paper birch. J. Org. Chem. 46: 1576-1578.

33. WALSKI, T.W., W.M. MAUTZ. 1977. Nutritional evaluation of three winter browse species of snowshoe hares. J. Wildl. Manag. 41: 144-146.

34. SCHWARTZ, C.C., A.W. FRANZMANN, D.C. JOHNSON. 1981. Moose Research Center Report. Alaska Dept. of Fish and Game. Vol. XII. Project progress report federal

aid in wildlife restoration project. W-21-2, Job
1.28R: 16-17.

35. OWEN-SMITH, N., P. NOVELLIE. 1982. What should a
clever ungulate eat? Am. Nat. 19: 151-178.

36. PALO, T.R., A. PEHRSON, P. KNUTSSON. 1983. Can birch
phenolics be of importance in the defense against
browsing vertebrates? Finn. Game Res. (in press).

37. PALO, T.R., P.G. KNUTSSON, K.H. KIESSLING. 1983.
Seasonal variations in ruminant in vitro digestibility of Betula pendula in relation to nutritional
content and phenolic constituents. Abstract from
the Third European Ecological Symposium on Animal-
Plant Interactions. August 22-26, 1983, Lund,
Sweden.

38. PALO, T.R. 1984. Distribution of birch (Betula spp.),
willow (Salix spp.) and poplar (Populus spp.)
secondary metabolites and their potential role as
chemical defense against herbivores. J. Chem.
Ecol. 10: 499-520.

39. FRYKHOLM, K.O. 1945. Bacteriological studies of
pinosylvin, its monomethyl and dimethyl ethers and
toxicological studies of pinosylvin. Nature 155:
454.

40. LYR, H. 1962. Detoxification of heartwood toxins and
chlorophenols by higher fungi. Nature 195: 289-290.

41. GRIME, J.P. 1977. Evidence for the existence of
three primary adaptive strategies in plants and its
relevance to ecological and evolutionary theory.
Am. Nat. 111: 1169-1194.

42. CHAPIN, F.S., III. 1980. The mineral nutrition of
wild plants. Annu. Rev. Ecol. Syst. 11: 233-260.

43. SMALL, E. 1972. Photosynthetic rates in relation to
nitrogen recycling as an adaptation to nutrient
deficiency in peat bog plants. Can. J. Bot. 50:
2227-2233.

44. COLEY, P.D. 1983. Herbivory and defensive characteristics of tree species in a lowland tropical forest.
Ecol. Monogr. 5: 209-233.

45. LEVITT, J. 1972. Responses of plants to environmental stresses. Academic Press, New York.

46. MOULD, E.C., C.T. ROBBINS. 1981. Evaluation of deter-
gent analysis in estimating nutritional value of
browse. J. Wildl. Manag. 45: 937-947.

47. MOULD, E.C., C.T. ROBBINS. 1982. Digestive capabili-
ties in elk compared to white-tailed deer. J.
Wildl. Manag. 46: 22-29.

48. FEENY, P. 1976. Plant apparency and chemical defense. In Biochemical Interactions Between Plants and Insects. (J.W. Wallace, R.L. Mansell, eds.), Plenum, New York, pp. 1-40.

49. RHOADES, D.F., R.G. CATES. 1976. Toward a general theory of plant antiherbivore chemistry. In J.W. Wallace, R.L. Mansell, eds., op. cit. Reference 48, pp. 168-213.

50. RHOADES, D.G. 1979. Evolution of plant chemical defense against herbivores. In Herbivores: Their Interactions With Secondary Plant Metabolites. (G.A. Rosenthall, D.H. Janzen, eds.), Academic Press, New York, pp. 4-48.

51. KOZLOWSKI, T.T. 1971. Growth and Development of Trees. Vol. 1, Academic Press, New York, 443 pp.

52. KRAMER, P.J., T.T. KOZLOWSKI. 1979. Physiology of Woody Plants. Academic Press, New York, 806 pp.

53. GARRISON, G.A. 1972. Carbohydrate reserves and response to use. In Wildland Shrubs - Their Biology and Utilization. U.S.D.A. For. Serv. Gen. Tech. Rept. in T-1. Utah State Univ. Press, Logan, Utah, 494 pp.

54. MOONEY, H.A., S.L. GULMON. 1982. Constraints on leaf structure and function in reference to herbivory. Bioscience 32: 198-206.

55. ALDOUS, C.M., S.E. ALDOUS. 1944. The snowshoe hare - a serious enemy of forest plantations. J. For. 42: 88-94.

56. GRANGE, W.B. 1949. The way to game abundance. Charles Scribner's Sons, New York, 365 pp.

57. GRANGE, W.B. 1965. Fire and tree growth relationships to snowshoe rabbits. In Proceedings of the Annual Tall Timbers Fire Ecology Conference 4: 110-125.

58. DODDS, D.G. 1960. Food competition and range relationships of moose and snowshoe hare in Newfoundland. J. Wildl. Manag. 24: 53-60.

59. DeVOS, A. 1964. Food utilization of snowshoe hares on Manitoulin Islands, Ontario. J. For. 62: 238-244.

60. PEASE, J.L., R.H. VOWLES, L.B. KEITH. 1979. Interaction of snowshoe hares and woody vegetation. J. Wildl. Manag. 43: 43-60.

61. WOLFF, J.O. 1980. Moose-snowshoe hare competition during peak hare densities. Proc. North American Moore Conference 16: 238-254.

62. WOLFF, J.O. 1980. The role of habitat patchiness in the population dynamics of snowshoe hares. Ecol. Monogr. 50: 111-129.
63. FISHER, R.A. 1958. The Genetical Theory of Natural Selection. 2nd Rev. Ed. Dover, New York, 291 pp.
64. ROUGHGARDEN, J. 1979. Theory of Population Genetics and Evolutionary Ecology: An Introduction. MacMillan, New York, 634 pp.
65. ZIMMERMANN, M.H., C.L. BROWN. 1971. Trees: Structure and Function. Springer-Verlag, New York, 336 pp.
66. ZIMMERMAN, R.H. 1984. Hormonal aspects of phase change. In Encyclopedia of Plant Physiology, Vol. 11. (R.P. Pharis, P.M. Reid, eds.), Springer-Verlag, New York. (in press).
67. HEIBERG, S.O., D.P. WHITE. 1951. Potassium deficiency of reforested pine and spruce stands in northern New York. Proc. Soil Sci. Soc. Am 15: 309-376.
68. LOYTTYNIEME, K. 1981. Nitrogen fertilization and nutrient contents in Scots pine in relation to the browsing preference by moose (Alces alces). Folia Forestalia 487: 1-14.
69. SULLIVAN, T.P., D.S. SULLIVAN. 1982. Influence of fertilization on feeding attacks to lodgepole pine by snowshoe hares and red squirrels. Forestry Chronicle, December: 263-266.
70. ROUSI, M. 1983. Susceptibility of pine to mammalian herbivores in northern Finland. Silva Fennica 17: 301-312.
71. WILDE, S.A., O.G. NULBANDOV, T.M. YU. 1948. Ash, protein and organo-solubles of jack-pine seedlings in relation to soil fertility. J. For. 46: 829-831.
72. McKEY, D. 1979. The distribution of secondary compounds within plants. In G.A. Rosenthal, D.H. Janzen, eds., op. cit. Reference 50, pp. 55-133.

DATE DUE